T0279835

# Into the Great Wide Ocean

## Life in the Least Known Habitat on Earth

Sönke Johnsen

With Illustrations by
Marlin Peterson

PRINCETON UNIVERSITY PRESS
PRINCETON & OXFORD

Copyright © 2024 by Princeton University Press

Princeton University Press is committed to the protection of copyright and the intellectual property our authors entrust to us. Copyright promotes the progress and integrity of knowledge. Thank you for supporting free speech and the global exchange of ideas by purchasing an authorized edition of this book. If you wish to reproduce or distribute any part of it in any form, please obtain permission.

Requests for permission to reproduce material from this work should be sent to permissions@press.princeton.edu

Published by Princeton University Press
41 William Street, Princeton, New Jersey 08540
99 Banbury Road, Oxford OX2 6JX

press.princeton.edu

All Rights Reserved

ISBN 978-0-691-18174-5
ISBN (e-book) 978-0-691-26415-8

British Library Cataloging-in-Publication Data is available

Editorial: Alison Kalett and Hallie Schaeffer
Production Editorial: Jenny Wolkowicki
Jacket and text design: Katie Osborne
Production: Danielle Amatucci
Publicity: Matthew Taylor and Kate Farquhar-Thomson
Copyeditor: Maia Vaswani

Jacket illustrations by Marlin Peterson

This book has been composed in Arno Pro

Printed in the United States of America

10 9 8 7 6 5 4 3 2 1

For Lynn
The Calvin to my Hobbes

# CONTENTS

# PREFACE

> If you're going to publish a book, you probably are going to make a fool of yourself.
>
> —Annie Dillard, interview in the *Chicago Tribune*, November 7, 2001

Like many marine biologists, I spend at most a few weeks a year at sea. Unlike most marine biologists, however, I don't even live on the coast. I wrote this book on a sunny porch, shared with my yellow Labrador retriever, Bonnie, overlooking our horse farm in central North Carolina. We have pastures, woods, a big red barn, and a little red barn, but the only water is the tadpole-infested pond that Bonnie so loves to swim in. The ocean itself is over two hours down the interstate.

That said, I feel so lucky. Not to be an academic, truth be told, the life of a professor is as full of bureaucracy, frustration, and tedium as the lives of many others. I feel lucky that—among the difficulties, indignities, and occasional crushing sadness of life—I have found both a world to love and people to love within it. The natural world, both on land and at sea, is beautiful, fascinating, and hilarious, and repays a life of devotion and curiosity. Nature is also right in a way that nothing else is, even under the pressures it now faces. Walk through a virgin hemlock forest or dive in the Sargasso Sea, and every speck of dirt on every mushroom, every sparkly yellow alga set against the blue water, is exactly where it should be, for now and always. Who can say that about anything else in our lives? Our messy, imperfect lives do have friends, though, and I am grateful for mine, and in particular for how they have helped me with this book.

First, I want to thank my editor, Alison Kalett. Authors often say that without their editor their book would not have been written. In my case, it's entirely true. I was years past signing the contract for this book, and over a year past the deadline for handing it in, but had not written a word. I would have bailed entirely, but I had used the advance to buy a tractor for the farm. Alison learned that I was coming to Princeton for another reason and pinned me down on a rainy November evening walk to her office. She gave me some free books, after which I felt I owed her one back. My mind works this way. Also, I love my tractor and don't want to give it up.

I'd also like to thank my lab members, colleagues, and various ocean-going crazies for helping me with this book. In particular, Alison, Anela, Brad, Cody, Dan, Edie, Eleanor, Emily, HBG, Heather, Jesse, Joe, Karen, Katie, Lori, Lynn, Nathan, Olivia, Sarah, Steve, and Tracey were generous with their comments, edits, and thoughts. Marlin Peterson drew the beautiful illustrations based on photos by Brad, Edie, Karen, Larry, Laura, Steve, and Tammy. Finally, I'd like to thank my wife, Lynn, my biggest fan and my harshest critic, for so many things, but especially for dragging me out to this farm and reminding me that it's never too late to have another adventure. And my daughter Zoë for being the coolest person I know.

# INTO THE GREAT WIDE OCEAN

## CHAPTER 1

# I Never Knew the Ocean

> Most of us walk unseeing through the world, unaware alike of its beauties, its wonders, and the strange and sometimes terrible intensity of the lives that are being lived about us.
>
> —Rachel Carson, *Silent Spring*

I thought I knew the ocean. At first, the ocean was an escape from the mills and pollution of 1970s Pittsburgh. Not that Pittsburgh was all bad. We had dinosaur bones in the museum, paddlewheel ships in the rivers, and red, inclined trams on the hills. I even enjoyed the "fires of hell" view of the blast furnaces of the Jones & Laughlin steel works that we'd pass on evening drives home from the mall. But the air had a brown edge, the older buildings were black from pollution, and although we did have parks, the trees in them often seemed to be fighting for their lives, pushing their roots through cracked concrete and gravel. My brother and I made the best of it, digging coal mines in our backyard and going on Cub Scout field trips, not to campgrounds but to factories. I loved and still love my hometown, but the unborn biologist in me lived for late summer when I could escape to a natural world that was happily alive.

Like millions before and after us, for two weeks each August, my family would go to Kill Devil Hills on North Carolina's Outer Banks. After a panicked early morning of my dad saying we couldn't possibly put another item in the trunk of the Dodge, we'd drive through the truck stops of Breezewood, Pennsylvania, around the DC beltline, past the cigarette-shaped headquarters of R. J. Reynolds Tobacco, finally stopping to get peanuts and a fried chicken dinner at the Virginia Diner in Wakefield and to sleep in a motel that—if the fates smiled on us—had a water slide. The next morning we'd shoot past the North Carolina farm towns of Coinjock, Jarvisburg, and Point Harbor, tiny places remembered by thousands of kids because they were the last countdown to the Wright Memorial Bridge that took you across the Currituck Sound, delivering you from an artificial world of bricks and sidewalks to one that mattered.

Having little money, we stayed in what I thought must have been the cheapest hotel on the Outer Banks, a "motor court" consisting of about twelve concrete, single-family structures, which would have looked like public toilets if they weren't painted pink, and had screened porches. Squat and lacking air-conditioning, they were cramped and oven-like, but they put me within a few dozen feet of a world that charted the course of my life for the next fifty years.

Many love the beach, but for me—raised by a lapsed Catholic and an atheist—the ocean immediately became the object of my awe and devotion. For those two weeks, I would spend nearly every waking moment in the water, splashing in the surf when in grade school, moving farther out as I grew, and eventually risking myself in larger and larger waves to see if the sea would finally swallow me up. I'd also talk to the ocean and send heartfelt goodbyes from the dunes when we had to leave. I was a lonely child, and the ocean provided a link to something powerful and beautiful.

And I thought I knew the ocean. It's true that barrier island beaches are simple, being more or less sand, water, and air, with sea oats and ghost crabs in the sand, a few unseen blue crabs and fish in the water, and seabirds in the air. Pulverized snails, clams, and jellyfish were tossed on the beach, along with some wood and the occasional horseshoe crab, but, overall, one of the charms of a North Carolina beach is its simplicity. However, I thought the beach was the ocean; that somehow the whole ocean was the sound of breaking waves, laughing gulls, and greenish murky water that smelled faintly of rotting seafood. I loved it more than I ever loved anything, but I was mistaking the peel for the apple, the bark for the tree, the skin for the person inside.

However limited in concept, my love for the ocean eventually propelled me to graduate school in marine biology. Coming from a nomadic life of teaching kindergartners, programming computers, and renovating houses, I had no training in biology. I also had no real appreciation for it, and chose the University of North Carolina mostly because I thought Chapel Hill was closer to Kill Devil Hills than it turned out to be. I suppose I was following in the footsteps of my mother, who went to the University of Kiel in Germany solely because it was near the beach. I did learn biology, and more importantly learned to love it, from Bill Kier, a kind and patient mentor. After two years, I had learned to scuba dive and was working on the behavior of brittle stars, which can be thought of either as speedy starfish or as five little snakes tied together by their heads. Being years behind everyone else in my graduate program, I developed a high-risk, low-reward project on vision in these animals, which never worked but did get me back into the water. I was again spending whole days immersed, but now snorkeling and diving on reef systems in the Florida Keys. This created my second false impression of what the ocean was.

For graduate-school me, the ocean was an oversized aquarium, clear and packed with life. The animals were mostly familiar—snails, clams, crabs, shrimp, sponges, and such—and mostly crawling on the bottom. I learned about these animals—who they were, how they were related, and how they made their way through life. And again, though not formally religious, I was in awe of the diversity and complexity of marine life. If the beach was the peel of the ocean, though, I was still only in the rind, the thin strip of shallow coastal waters that skirts the sea. I happily explored this rind for six years, not stopping to think what might be beyond it.

As for so many people, my life then turned on a chance moment, in this case a conversation with Bill at the end of my graduate training. As I mentioned before, my graduate project was best forgotten. So, I needed a new and hopefully better project to continue the next step of my training, as what is called a "postdoc." Being interested in eyes and optics, I thought about studying human lens cataracts. Bill, always gentle with his words, suggested that just maybe a lot had already been done on this topic (in fact there are many books and hundreds of articles), but that if I was interested in the transparency of lenses, perhaps I'd be interested in the transparency of entire oceanic animals. I still remember exactly where I was standing in his office, with its barred windows looking out on the Geology Department's rock garden. "There are so many transparent animals?" "Yes," he answered, "and many of them are clear as glass. The animals out there look nothing like the animals on land or near shore." In that moment everything in my life changed, as I became aware of an immense world that had always been right there, waiting for me to notice.

I did know the open ocean existed, of course, perhaps more than many, because I had crossed it. I knew that, as a baby, I had emigrated to America on an ocean liner, in a pillow-padded orange crate nailed to the deck in my parents' cabin. In the back of my mind, I'd always wanted to go back to sea, to be out of sight of land again on a real ship. But I had no idea that the waters beyond sight of land held animals different from those I saw swimming at the edges of coral reefs. I expected whales, sharks, and fish, with maybe the occasional jellyfish floating by. Over the next ten minutes or so, though, Bill pulled down books from his own time as a postdoc at Woods Hole Oceanographic Institution, showing me pictures of animals that I didn't even know existed, or even could exist.

I had to go. I applied to get a postdoc at Woods Hole or Harbor Branch Oceanographic Institution in Florida, assuming that at least one would appreciate my enthusiasm and overlook my lack of knowledge. Both turned me down flat. I cleaned aquaria for the next year in the teaching labs at UNC and applied again. This time, they both accepted me, each allowing me to do two eighteen-month fellowships one after the other, giving me three years at places where ships and submarines explored the open sea.

I chose to go to Harbor Branch first, arriving in a crammed rented truck, followed by my wife in our hatchback with the stereo, dog, and cats, and started working with Edie Widder, who was so passionate about the various ways that animals emit light that she would answer the phone saying, "bioluminescence." After a quick tour of her lab (painted black floor to ceiling for better optical measurements), followed by gin and tonics at her sound-front home, I asked her when we'd be going out to sea. She mentioned that the *Edwin Link* would be coming back into port in a few weeks, and we would be sailing on her next. I was immediately impressed that we would be using a ship large enough to have a name, but did my best to act nonchalant.

Harbor Branch Oceanographic was built around an artificial harbor, and my office had enormous windows that overlooked this channel. Each day I worked on my computer, peeking over the monitor now and again to see if the *Edwin Link* was there. One day I heard a deep humming and saw the tip of the bow at the right side of the window. The ship was white, and like the Star Destroyer in the opening moments of *Star Wars*, it kept getting bigger and bigger as more and more of it entered view. And, just like when I was an eleven-year-old boy in the theater in 1977 watching that iconic opening scene, I was more excited than I thought possible. I remember thinking: *This is a ship. Something that is large and strong enough to take us hundreds of miles offshore, for months if needed. It is exploration incarnate.* In truth, the RV *Edwin Link* was only a middling-sized research vessel (about 170 feet long), a converted fishing boat in fact, but at that moment I felt like I was looking at a Saturn V rocket that was preparing to take me to Mars.

A week or so later, after a dirty and sweaty day of loading tubs of nets, cases of deep-sea instruments, tanks of compressed air, and what seemed to be an unnecessary number of boxy old-school monitors, we all hugged our families on the dock and pushed off. After a short passage through the calm, brown Indian River Lagoon, we turned through an inlet straight into the waves of the open sea to begin a four-day trip up the Gulf Stream to Cape Cod, where we would be working for two weeks. It was there that my imagined voyage to the unknown on the *Exploration Incarnate* met the realities of being offshore for the first time.

We had rough weather, meaning high wind. The Gulf Stream, a large river within the ocean, can also be a wavy place. Together, the wind and current built up waves of 12–15 feet, and—aside from my orange-crate trip to the States—I had spent almost no time offshore. It's an old saying that it's seasickness when you're afraid you might die, but bad seasickness when you're afraid you might not. I was well into the latter stage, and able to keep my food down only by sitting outside at the very middle of the ship (which rocks the least) and locking my eyes on the horizon. Unfortunately, this middle location was exposed, and periodically a large wave would hit me. I soon learned that shoes soaked in salt water are never the same again. My spot also put me in the path of the nauseating diesel fumes from the smokestack. And, because bad things come in threes, I had also gotten unlucky with food; the cook had brought about fifteen drums of "Nutrafat" on board, using it to cook nearly all our meals, which rolled around my plate in a queasy manner. Edie called it my "trial by fire," and the seasickness, along with other privations and indignities (try showering when you can't stand), wore me down. But I was there, on the ocean, out of sight of land, exploring it with nets, instruments, and a manned submersible. It was then I learned that I never knew the ocean at all.

## The Great Wide Open

The world I saw over the next two weeks was not only unlike the beach and the coral reefs I had played and worked on, it was unlike any habitat I had ever seen or imagined. The water itself was so blue it looked fake. Before graduate school, I had hitchhiked in Oregon and seen Crater Lake. That was the only other time I had seen water like that, looking as though it had been painted. It was so uniformly and deeply blue that it didn't look like a transparent substance at all, so it shocked me to see a fish and squid far below the surface. I didn't get to scuba dive or go down in the submersible on that first cruise, but when I did a few months later, the blue color below the surface was even more intense, becoming paler as one looked up and deepening to purple as one looked down. The subsurface ocean was also clearer than a swimming pool, and I could easily see large fish, other divers, and even the bottom of our dive boat from over 100 feet away.

Unlike the ocean of all my other experiences, it had no smell and was almost silent unless the wind was high enough to cause whitecaps. In fact, with the ship's engine off, life at sea on calm days is about as quiet as it gets, with only little slaps of water against the hull. Unlike the restless surf of the beach, which I so loved, the ocean below the surface was deathly still.

Above all, the ocean appeared empty. Growing up in Pittsburgh and then living and working in large cities and bustling university towns, my eyes, ears, and even body were used to the continual crush of activity and life. Even the desertlike beaches of North Carolina were full of things to pore over. But offshore, at first glance at least, there was almost nothing. No other ships, no planes, no birds, and—when seen from the deck—often no animals. Just the wake of the ship to let you know you were still going somewhere. This empty feeling was even stronger at night. My former student Julia was on a research cruise a couple years ago and said that those on the night shift would often feel as if there was a black velvet curtain around the ship, one that you could touch if you only reached a bit farther. This curtain appeared to enclose the entire world.

It was the animals, though, the ones that we brought up in the nets and the buckets of the submersible, and that I later saw directly while underwater, that truly surprised me. Aside from some of the fish and jellyfish, I had never seen anything like them, even though most were related to animals that I had studied in school. There were winged snails that flew through the water like birds, long worms paddling with dozens of crystal oars, and 50-foot-long chains of tubular animals, each pumping to both suck in food and move through the water. Every time we brought in the net or animals from the submersible, I would have to ask, "What on earth is that?" More experienced people would kindly tell me, only for me to often respond, "Are you serious?" In my defense, imagine if someone showed you a gelatinous ball with wings and a sofa-sized snot web that it used for a feeding net—and then told you it was a snail.

After a few cruises at Harbor Branch over the next year, two things became clear. First, that the animals that swim or drift in the water of the ocean, as opposed to living on the bottom, look odd to us because they have to solve the problems of life in a habitat we don't share. They need to stay at a given depth, not get crushed by pressure, move, find food, avoid becoming food, find one another, and make their way through their lives in a habitat with very different rules from those that govern life on land, or even life in coastal waters. They seem alien because the ocean is an alien world to us. The second realization was that this alien world *is* our planet, or at least the vast majority of it. About 90 percent of the earth's habitable space is in the water of the ocean. This is not only because the ocean covers more of the earth's surface than land does, but also because the ocean is much deeper than the range of height that is inhabited on land. The ocean is on average 2 miles deep, while most plants and animals live within a 300-foot-high strip along the surface of the land, with only some birds and insects going several hundred feet above that. We are a water planet, and the rules of the oceanic realm are the primary rules of life. Our slender terrestrial world, and the rules that govern it, are the exception. To know the ocean is to know our planet.

## What This Book Is About

This book is about the adaptations required to thrive in the pelagic portion of the ocean. By "pelagic," I mean everything that is not the bottom, which includes the surface and the watery world between it and the seafloor. This is also referred to as the "water column," the "midwater," or the "open ocean." While I will occasionally discuss life near the coast, I will mostly focus on the pelagic world that is found far from the shore. It is here that we find adaptations that appear strange to us. By adaptation, I mean the results of evolution via natural selection, where natural selection is the process by which organisms within a species that are better able to live and reproduce become more abundant. This process, which has been termed "climbing mount improbable" by Richard Dawkins, has led in this case to the wondrous forms we see in the pelagic.

This book has two interwoven stories. One is about life in the open sea, particularly in the top 1,000 feet, where there is still a reasonable amount of light (just enough to read this book by). In it, I discuss how pelagic animals, often so strange to us, manage to solve the problems of this world. The first section focuses on the physical issues that these animals face. Chapter 2 discusses the various ways in which the animals keep from sinking to the bottom. Chapters 3 and 4 delve into pressure and light, the first factor going up rapidly with depth, and the second going down even more rapidly. The physics section ends with chapter 5, which explores the various creative ways in which animals move through water and discusses what is known as "daily vertical migration"—the largest movement of animals on our planet. The rest of the book deals with the biological aspects of this oceanic world. Chapter 6 talks about how animals feed while managing to avoid becoming food themselves. Chapter 7 tries to answer the question of how the residents of the vast and mostly empty sea manage to find one another to reproduce, especially when most are already so camouflaged that they are almost imperceptible. In chapter 8 we end the section with a discussion of communities in the ocean, looking at relationships both within and between species.

Within the story of the animals is the second story of this book, that of the people who study them. Trying to understand a world that is hundreds of miles from land and a thousand feet below you can be frustrating and made worse by the fact that using ships is enormously expensive, typically $15,000 to $50,000 per day. Imagine wanting to answer a question and being told that you need to buy a brand new SUV every day for several weeks, and that on some of those days you won't be able to do anything because it's windy or because some piece of machinery is broken and that you still have to pay for the next car. Imagine a hurricane coming, and you being left with nothing but the bill for a couple dozen SUVs.

At sea, plan A never works, and plan B is usually a pipe dream, so you're left with plan C or—more often—plan D, which you built out of old ladders, spackle tubs, and duct tape on the back deck at three a.m. Add in weather, seasickness, and sleeping in a 5 foot 10 bed when you are 6 foot 4, and it all can be a trial. However, there is also camaraderie, seeing things that no one else has seen, and the heart-stopping thrill when something . . . actually . . . works. I like to tell my students that everyone wants to go to sea once, but far fewer ever want to go again. Those few are hooked for life, though, to be honest, they're crazy. Sarah, a previous postdoc, told me that going to sea was perfect for people who would be happily institutionalized, and Edie once said that offshore life was a week of boredom punctuated by five minutes of rabid excitement. Like the animals, human oceanographers have adapted to an alien environment.

Along with the story of the researchers, this book is also about the technology required to study pelagic habitat. Pressure, gravity, light, and motion are just as important concerns for our equipment as they are for the animals, so specialized techniques are required. Some methods, such as trawling, are quite old, but have been modified so that the animals a mile down can be collected in healthy condition. Others, such as scuba, are more recent, but have also been heavily modified for safe work in the open ocean. There are also the high-tech instruments: the manned submersibles, the unoccupied submersibles (ROVs), and the self-driving AUVs, all built to survive pressures that would crush a car into a cube. The equipment we deploy is often only half the story, because we typically want to work on the animals we collect. We want to understand their senses, how their bodies work, and possibly even their behavior. To do this, we design cramped and chaotic home brew labs, usually brought aboard in multiple plastic tubs and rapidly assembled and tied down right before the ship leaves the dock.

Overall, I want to tell the story of a world that almost no one sees but that nevertheless dominates our planet, and to show how the animals and the researchers manage to survive and thrive within it. I myself will never really know the ocean. Like anything—or anyone—there is always more to learn. On my last cruise we filmed a giant squid using a camera on a mile of rope that we pulled in by hand every night. There's always something new.

## What This Book Is Not About

This is not a book about the deep sea. The simple reason for this is that there are already so many excellent books on this subject. The top 1,000 feet of the ocean has received far less attention, despite its being where most of the life is. It's also remarkably beautiful.

This book is also not about the bottom of the sea. This is for a few reasons. First, the animals at the bottom are not nearly as alien as those in the water column. A snail at the bottom at 1,000 feet still looks pretty much like a snail you'd see on a coral reef, but a snail in the water column is unrecognizable. Second, many of the challenges in the ocean—buoyancy, finding one another—are most relevant in the water column.

I am also not writing about algae, bacteria, protists, or viruses. Algae are the primary food source in the oceanic water column, and their consumption and eventual transport to the bottom as fecal pellets are important for removing carbon dioxide from the air. I am not a botanist or microbiologist, though, and am following the rule, "write what you know." Maybe that's not always the best advice, but I think it's wise for science writing.

Finally, this is not directly a book about conservation. This is not because I don't care about it. Like many biologists, nature is my religion in a visceral way. Nine years ago, I was in Pisa for a conference, during which our family dog Bandit died of cancer back in North Carolina. Distraught, I went into the Pisa Cathedral (next to the leaning tower), and—despite my lack of religious upbringing—lit a candle for my dog and sat down in a pew. Thoughts of my pet developed and grew over the next ten minutes, until eventually I had an internal vision of life on our planet, with pink jellyfish moving just below the surface of the ocean, shiny green tiger beetles crawling through leaf litter, and fat alligators lying at the edge of Florida swamps, what felt for a moment like all life on earth moving together in an endlessly complex and endlessly evolving dance. This feeling had never been so strong before, but I've felt it my entire life. Our natural world and our ability to appreciate it are stunning gifts, more than we deserve, and never to be cast aside.

We save what we love, and we love what we see. The world below the surface of the open sea has been seen by only a tiny fraction of people. My guess is that only a couple of hundred have gone down in submersibles and perhaps only a few dozen have scuba dived in the open ocean. Before we as scientists can ask people to preserve this important and fragile habitat, we need to show them that it's there and the beauty of what lives in it. To borrow from Doctor Seuss: like Horton the elephant, I need to convey a message from the inhabitants of this hidden world—"We are here."

# CHAPTER 2

# Gravity

More weight.

—Last words of Giles Corey at his execution, following the Salem witch trials, as dramatized in Arthur Miller's play *The Crucible*

## Drowning

A research vessel is many things. Outside, it is a floating factory, with winches, cranes, A-frames, and other hunks of moving steel that a terrestrial scientist seldom sees, let alone works with. Inside, it is a laboratory that seagoing scientists fill with their equipment and supplies at the start of the cruise, usually during a frantic, sweaty "mobilization" day, when delicate and expensive equipment in dozens of crates is dragged onto the ship, unpacked, set up on preexisting counters, and tied down with bungee cords so that it doesn't fall off when the boat rolls. Inside, it is also a communal dorm, typically with tiny rooms, a kitchen and a dining room, a machine shop, and possibly a TV lounge, an exercise room, or a limited hospital. Above all this sits the bridge, with its nest of navigational and control equipment, crowned with a small forest of antennae, satellite dishes, and forever-twirling radar wands. Below, like a double bass, sits the enormous engine, whose sounds tell an experienced person lying in their bunk everything they need to know about what the ship is doing or about to do. In sum, a research vessel is a hybrid of a commercial trawler, a university lab, and an experiment in group living, with enough food and fuel to survive for months wherever you choose to go. Underlying it all, though, is this root truth—out there, a research vessel is the only thing between you and drowning.

At the start of every cruise, usually an hour after the ship has left the port, we have a safety briefing. The alarm bell goes off, and we all put on our life jackets, grab our fluorescent neoprene survival suits, and go to the muster point. The meeting is usually led by the first mate and lasts about an hour. Topics discussed range from which lifeboat you will run to if the ship goes down in flames, to the Wi-Fi password, to how unhappy the engineer will be if you flush anything novel down the skinny, vacuum-controlled sewer tubes. Nearly always it ends with the first mate asking some poor new person to fight their way into their survival suit while their friends take videos. Most importantly, we discuss the various ways one can become injured on a ship. Some of these are uncommon in scientific life and, given the heavy machinery involved, quite gruesome, but the worst thing that can happen is for someone to fall overboard.

At first, this seems like an exaggerated fear. Most oceanic research vessels don't go much more than 10 miles an hour, which is about as fast as people drive in parking lots. Many times we are going much more slowly, perhaps pulling a net at a mile an hour. We may be leaning over the side of the ship during these trawls, talking and watching the jellyfish go by. I've done this for countless hours. The water is only 6 feet away, you can almost touch it, but I know that if I fall in, I will likely die. As Edie told me on my first cruise, "Assume you are walking along a 3,000-foot cliff—that rolls." My friend Tracey Sutton likes to say that if you fall off a ship at night, your chances of recovery are one in seventy-five thousand.

So what's the problem? After all, if you fell out of a dune buggy in an empty desert, even at night, as long as someone saw you drop they'd just turn around the car and get you. At sea, there are factors that make this form of recovery almost impossible. First, large ships are hard to turn around. For example, the research vessel *Atlantis* from Woods Hole Oceanographic Institution weighs about 3,000 tons, about two thousand times the weight of my car. It takes time to get something that heavy moving, so it takes time and distance to get it turned around. If a person falls off a ship going 10 miles an hour, even if the ship immediately starts turning, it might take fifteen to thirty minutes for it to slow down, turn around, and come back to the same spot at a safe speed. The second problem is that the ship may not immediately turn. Even if someone is next to me on the back deck when I fall in, they have to either run to the bridge or find a radio to contact someone in charge. During this time, I will have drifted out of sight. Which brings me to the biggest problem: even during the day in sunny clear weather, it's almost impossible to spot something in the water. Humans are just barely buoyant (more on that later), so at best only their head is sticking out of the water. The open ocean, while not usually the churning disaster shown in movies, is seldom flat. The waves are seldom less than 2 to 4 feet high, and usually at least 6 feet high. So, once someone drifts about 50 feet away from the boat, they're nearly impossible to find. In fact, if someone goes overboard, we're told to not run to get help, but instead to keep our eyes on the person, throw everything we can find into the water to make a marker, and scream like the world is ending until someone else figures out what's going on and contacts the bridge.

All that said, I do know one person who fell off a ship and survived. My friend Brad Seibel, a deep-sea squid biologist, and his postdoc were catching squid over the side of the boat while both were in their flip-flops. Brad did not fall in, but his postdoc did. The postdoc was lucky in that he was among many people, the boat was stationary, and the sea was flat calm. The harsh truth, though, is that if you fall off a moving ship—even during the day, and God forbid at night—you will drown. If you fall into cold water, your limbs will quickly become useless, perhaps in only fifteen minutes, and you'll drown. If you fall into warm water, you may be able to tread water long enough to attract oceanic sharks, who will eventually attack you, and you'll drown. Another victim of gravity, the death hug of the planet.

Like so many of the common experiences of our life, gravity only gets odder the longer you look at it. The usual explanation is that gravity is the force that pulls you down. But this downward force is actually the sum of forces pulling you in all directions. At this moment, parts of the earth are indeed pulling you down, but other parts—for example, nearby hills—are pulling you sideways. The sun is pulling you up during the day, and the moon, stars, and planets are pulling you every which way depending on where they are. Your bathtub is pulling you to the bathroom, and your car is pulling you to the garage. Everything in the universe is pulling on everything—which is both comforting and unnerving. That the end result for us feels like a steady pull downward is because of one of those mathematical miracles that physicists like my dad love. The amount of pull from each object is proportional to the mass of the object and decreases with the square of the distance from the object (so if something is twice as far away it only pulls one quarter as hard). You may have heard of this law, known as the inverse square law, but the astonishing thing to me is that there is no universally accepted explanation for it. As my dad likes to say, physics just is. The laws of gravity do make life tidier. The pulls of the stars don't affect us because they're so far away, and the pulls of the sun and moon really show up only in our tides (and, of course, in keeping our planet going around the sun). The inverse square law also means that all the pulls from all the different parts of the earth, from its rocky crust to its iron core, add up to just one pull coming from the center of the earth.

In daily life, gravity is convenient. It keeps us in our seats, lets us walk, and makes sure that when I toss a ball for my dog Bonnie, she'll be able to catch it. It does make our feet, knees, and backs sore after long days (and eventually after long years), but without gravity our life would be tricky. It's only when we fall that gravity shows its teeth. Or when we fall into water. Because, while falling into water will not break our bones, gravity while we are submerged dramatically increases the amount of energy we need to expend to stay alive.

When my brother and I were in grade school, we spent our summers at a day camp run by the university where my dad worked. The camp was at the top of one of the tallest hills in Pittsburgh, so getting there was already a workout, but the swim portion of camp seemed to be run by people with unreasonably high expectations of the fitness of eight-year-olds. Each class started with us treading water in the 16-foot-deep end of the Olympic-sized pool. During the first fifteen minutes, we were allowed to use our hands, but for the next fifteen we had to hold our hands above our heads. This not only left our legs doing all the work but added the weight of our arms, which were now out of water. By the end of the summer, we had to do the whole thirty minutes with our hands out of the water. I tried this and threw up; vomit is not easy to clean in a pool that size, so they didn't ask me to do it again. So, when I started doing open-ocean scuba diving, the first thing I thought (after "Please don't let me see a shark") was, "How do these animals keep from sinking?" and then, "For those that have to swim upward their whole lives, how can they stand it?" It turns out that pelagic animals have found a number of ingenious ways to solve the problem of gravity, which we will get to, but first we need to talk about buoyancy.

## Scuba Diving and the Principles of Buoyancy

Like many scuba divers, I first dove only on coral reefs and shipwrecks, where the ground was not far below. Now and then, I might have been on a wall reef, but even with these there was always something solid that I could hold on to in an emergency, even if it was not directly below but instead in front of me. It wasn't until I started doing what is known as blue-water diving that I started to think about buoyancy. This sort of diving is just like any other, except that the bottom may be miles below you. Seems trivial, but it matters. During one dive, I handed my bag of carefully collected specimens to a fellow diver. He promptly dropped it. I remember watching it slowly go down and then glaring at him. We finished the dive, got back into the small zodiac boat, drove it back to the ship, took apart and cleaned our dive gear, took off and cleaned our wet suits, and went to take showers. Soon after, we were eating dinner, damp and tired, and—while I was chasing a potato that was rolling around my plate—it occurred to me that my dive bag was still going down.

So, when diving this far offshore, we are tied to a tether system of thin ropes that are connected to a thicker rope that is tied to both a float and a small boat. These tethers are custom-built by the divers themselves, and building one is a rite of passage.

Mine is made of climbing ropes of different colors, toilet parts, and a host of brass clips. The purpose of this system is not to support our weight, but to keep us from drifting away in the featureless open sea. Our weight is instead supported by a float that we wear like a vest, our wetsuit, and—to a lesser extent—by the air in our own lungs. All divers use these flotation vests, but depth control is typically easier for reef divers because they have a frame of reference, which is the reef itself. In blue-water diving, the only possible reference points are the other divers and the tether system, which are often moving. We often face away from this tether, which leaves us facing an expanse of blue nothing. So, the vests—known as BCDs, for "buoyancy control devices"—and our lungs are continually on our minds.

It's often said that Archimedes discovered the principle of buoyancy in his bath, shouting, "Eureka!" and possibly running naked through the streets of Syracuse. In fact, he instead discovered that the rise in the water level in his tub could tell him the volume of that part of his body that was under the water, which solved the problem of how to measure the volume of oddly shaped objects. This, along with an object's mass, could be used to determine its density, which in Archimedes's case allowed him to determine whether a crown was solid gold or not. Archimedes's principle of buoyancy—which, if it had a eureka moment associated with it, has been lost to history—is that the buoyant force of an object is equal to the weight of the water it displaces. This principle allows steel ships to float, makes scuba diving possible, and solves the gravity problem for many pelagic animals.

Our lungs displace about 1–3 quarts of water, depending on how inflated they are and how large we are. Water is also heavy, weighing about 2 pounds a quart. So, lungs give us about 2–6 pounds of buoyancy. You might think that's not nearly enough, but most of our weight is from our muscles and organs. These are only slightly denser than water and so don't pull us down much when they're in water. Our bones are about twice as dense as water, but our fat is less dense than water, which just about compensates for the bones and the slight negative buoyancy of the muscles and organs. Add in the buoyancy from the lungs and the average density of humans is very close to that of water. So why does anyone drown?

Indeed, many people can float on their backs with little effort, especially on salt water, which is denser than freshwater. In that summer camp I spoke of, we were taught the "survival float," which is floating on your stomach, lifting your head now and then to breathe. The problem is that because our average density is so close to that of seawater, it's strongly affected by lung capacity and fat percentage. The smaller your lungs relative to your body size, or the leaner you are, the harder it is to stay afloat. I was a rail-thin kid and could not float to save my life. Without continual kicking, I went straight down. As I entered my teens and added fat tissue, I grew more buoyant, but could stay afloat only if I fully inflated my lungs and held my breath. Now, as with most American adults, my body fat and muscle percentages are such that I float relatively easily. My positive buoyancy is still critically dependent on full lungs, though. Those who drown near shore often do so because an initial swallowing of water leads to a panic spiral of shallow breathing, which decreases buoyancy. Those who drown offshore, assuming they can swim, likely do so from hypothermia or from exhaustion in the wavy ocean. Oddly, for what appears to be a passive process, staying afloat as a human takes effort, because the legs—which have considerable muscle and bone and no air—are continually pulling one down. Once one drops below the surface, the water pressure begins to compress the lungs, reducing their volume and thus their buoyancy. This gets worse with depth, creating a positive feedback loop that increases the chances of drowning in an exhausted person.

This last statement touches on a principle that is fundamental to buoyancy in the ocean. Air compresses easily, water barely compresses at all. We will discuss this more in chapter 3, but water is surprisingly heavy and thus exerts substantial pressure, even at shallow depths. So, any person or animal using a nonrigid gas float as a buoyancy control device (BCD)has to deal with the fact that the float changes volume with depth. Scuba divers are trained endlessly on this principle, because failing to understand it leads to crises. A major issue is that buoyancy provided by nonrigid floats is unstable. Suppose I'm blue-water diving at 60 feet, using a BCD that's inflated with about 5 quarts of air. As I've mentioned, humans are roughly neutrally buoyant. Scuba divers also breathe air that has the same pressure as the water pressure at their depth. So their lungs are inflated to the same extent as they would be at the surface. The roughly 10 pounds of buoyancy created by the BCD is mostly offsetting the weight of the BCD itself, my other dive gear, and any collecting equipment I might have. If I've done it right I'm not going up or down, and I'm free to look around for animals to study or collect. But suppose I see an animal 10 feet above me and swim up to collect it? The water pressure at 50 feet is less than it was at 60 feet. The breathing regulator has compensated for this, and my lungs are still inflated as if I were at the surface (I've always found this technical trick of scuba gear to be magical). The dive gear still weighs the same as it did, and its volume hasn't changed. However, the lower water pressure at 50 feet has let the air in my BCD expand. So, now it displaces more water, which means its buoyancy has gone up. Thus, my carefully negotiated neutral buoyancy has switched to positive buoyancy.

If I don't pay attention at this point, I may be pulled up to 40 feet, at which the BCD will be even larger, giving me even more buoyancy and greater upward speed. If I'm not careful, I could rapidly ascend to the surface, leading to a number of serious conditions, including neurological damage and exploded lungs, which will be discussed in chapter 3. This is why the first thing that scuba divers do at the end of their dive is to open the escape valve of their BCD to let this ever-expanding air escape as they rise. It's also why scuba divers are trained to exhale as they ascend so that compressed air doesn't expand in their lungs and possibly rupture them.

The same problem would occur if I saw an exciting animal 10 feet below me. The air in my BCD would now compress, reducing the amount of water it displaces and thus its buoyancy. Unless I corrected for this, I'd start going down faster and faster. The water in the open ocean is often so clear that it doesn't get darker rapidly with depth the way it does on a coral reef. I've been in submarines down to about 3,000 feet and have always been surprised at how bright it is even at 600 feet. So, dropping from 50 to 100 feet isn't noticeable if you're focused on your work. At first, this descent isn't as dangerous as an uncontrolled ascent, but by about 100 feet the nitrogen pressure in my lungs would be high enough to make me feel drunk, and by 150 feet the oxygen pressure would be high enough to start to become toxic (again, more on this in chapter 3). Breathing in and out also changes our depth. If you're neutrally buoyant and hold a deep breath, you'll start an uncontrolled ascent. If you exhale and keep your lungs relatively deflated, you'll start an uncontrolled descent. So the tether system we use has a second purpose—to keep us from going up and down by too much. It is frowned upon to use the tether for this purpose, though, because you're pulling everyone else up and down as well.

I'm belaboring all this to make it clear that gravity and buoyancy in the ocean can be a royal pain, especially for humans and animals that use nonrigid floats. That said, the number one impression I've gotten from blue-water diving is that of peace. My first blue-water dive was in 1998 in the Gulf of Mexico, a few hundred miles west of Tampa, from a smallish research vessel out of Louisiana called the *Pelican*. We launched a small zodiac boat from the ship, drove about a mile away, set up the tether system, tied into it, and dropped. I wish I had a video of the descent from above, because we were arranged in a wide circle around the tether, falling like slow-motion skydivers in a blue sky that had no earth below. The bottom was about 2 miles down and visibility through this striking blue water was at least 100 feet. We tied off at 60 feet, pulled glass jars from our netted collection bags, and began looking for animals. Soon it was obvious that we were surrounded by translucent beings of all kinds, shapes, and sizes, ranging from 30-foot-long trains of a colonial animal known as a salp to pinkish moon jellies that were a foot across. While a few smaller animals seemed to be fighting their way upward as I did at that summer camp, for most it was as if gravity didn't exist at all. As if we were in outer space, between the galaxies, but everything was lit with blue light. It was the most astonishing thing I'd ever seen, and its memory still sits with me like a first kiss or the birth of a child.

So how do these animals manage to make the most pervasive force in our existence, the one that has haunted sailors for thousands of years with the threat of a terrifying death, simply go away? As is nearly always true in biology, natural selection has resulted in a number of diverse solutions.

## Going Light

When I'm not working, I like to backpack. As anyone who backpacks can tell you, it's no fun to carry your life on your back like a snail. This has led to a minor industry in gear and tips for ultralight backpacking. The tents have become so light that you can see through them, and I've had friends cut their toothbrushes and forks in half to save that extra ounce. It gets to be a bit much, and people look ridiculous trying to use a 2-inch toothbrush, but it underlines the point that if you want to reduce the effects of gravity, the first thing to consider is dropping weight.

Many pelagic animals are good at this, reducing whatever heavy parts they have and expanding other parts that are mostly water and thus close to neutrally buoyant. Some of my favorites are the pelagic snails, in particular the ones called heteropods (see figure 1) and the pseudothecosome pteropods. These animals are so adapted to a pelagic existence that in many cases it's hard to recognize them as snails at all. The heteropods look like miniature transparent elephants, and some of them carry small shells halfway down their bodies like mini-backpacks. The shell is still hard, but much thinner and thus lighter, and of course much smaller than you would usually see for a snail that size. The body of the animal itself is best described as gelatinized, in that the material appears to be the same as that found in many jellyfish. The internal organs are small and often crammed together in a mirrored pouch (more on that later). The foot of the heteropod has been modified into a sculling fin, which it uses to move about. The rest of the body has little muscle. So the animals have reduced or eliminated the heavier parts of their bodies and replaced them with gelatinous tissue. The pseudothecosome pteropods are similar, but their foot is divided into two large wings that they flap to fly underwater, and the

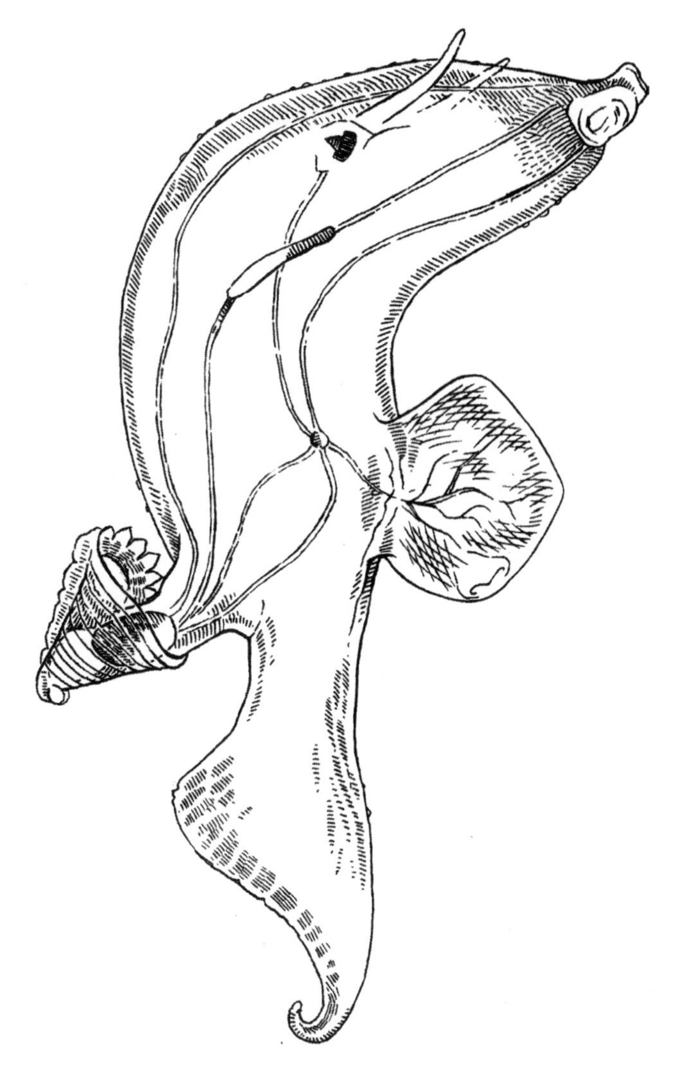

FIGURE 1. The shelled heteropod *Carinaria*

"shell," while still large, is also gelatinous, making it both transparent and much lighter.

Other pelagic animals don't so much go light as go wide, spreading themselves out so that they sink more slowly. This works well in larvae (which we discuss more in chapter 7) because they are small. In small animals, the primary factor that affects the sinking rate is the amount of surface area. So, many of these larvae look like tiny Sputniks, with long protuberances in all directions. In a way, they're only postponing the inevitable, since they're still negatively buoyant and will sink, but sinking slowly is better than sinking quickly. Of course, these same appendages make it harder to swim back up, but some of the animals pull them together when rising, much like closing an umbrella.

In the end, though, as with love, you can give away only so much of yourself. And your appendages can be only so many or so long before things get unwieldy. Therefore, some animals resort to other tricks.

## The Surface Dwellers

Some pelagic animals avoid the fussier aspects of maintaining the correct depth by simply having so much flotation that they bob at the surface. Some of the best examples of these are the chondrophores and a few of the siphonophores. Siphonophores are a group of about two hundred species of gelatinous animals that are related to jellyfish but typically look quite different. Instead of a pulsating disk with trailing tentacles, they are typically composed of a large collection of what can appear to be individual gelatinous animals but are in fact different parts of one animal. These parts also come in only a few types. One type may consist of pulsating structures that move the whole animal along, another type may be responsible only for feeding, another only for reproduction. The whole thing appears like a colony, but with division of labor. In fact, these animals have fascinated biologists for decades because it seems that they live at the boundary of being a colony and being an individual with

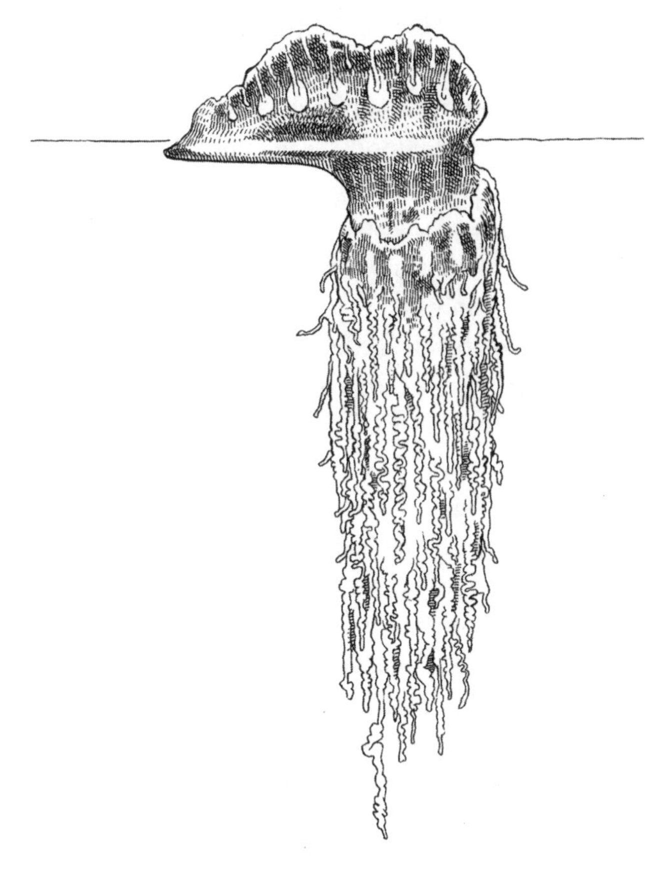

FIGURE 2. The Portuguese man-of-war, *Physalia physalis*

organs. Some are many feet long, and appear grotesque, like some overly tentacled horror from a sci-fi novel. It doesn't help that many give a painful sting. Perhaps the best known siphonophore is the Portuguese man-of-war (*Physalia physalis*; figure 2), which washes up on beaches and gives a nasty sting. Chondrophores are similar to siphonophores in their semicolonial, divided-labor aspect, and are more closely related to siphonophores than they are to jellyfish. They're a much smaller group, though, with only a handful of species, all of which float on the ocean surface.

Aside from its powerful sting, which it uses to kill small fish, shrimp, and plankton for food, the man-of-war is known for its pneumatophore, a large, translucent, blue-purple float that has always looked to me like a spat-out bubble gum bubble. Typically the float, which can be almost a foot long and up to 6 inches high, is the only thing a swimmer will see, and is a good warning to back off before the feet-long stinging tentacles make contact. Research suggests that the animal can partially deflate its float to dive below the surface, and then reinflate it at depth when it wishes to return. The fascinating aspect of this is that it does so by using a gland attached to the float that generates carbon monoxide from the amino acid L-serine. Carbon monoxide is best known to humans as a deadly gas formed by car engines and other forms of combustion. It binds irreversibly to the hemoglobin in our blood, preventing oxygen from doing so. In high enough concentrations, it leads to death by oxygen starvation in our tissues, even though our skin still looks cherry red. Carbon monoxide is otherwise rare in nature (but, oddly, it is also found in kelp floats), so it was a surprise to find it in man-of-war floats. Once the float returns to the surface, the carbon monoxide starts to diffuse out through its thin walls, and gases from the atmosphere (mostly oxygen, nitrogen, and argon) diffuse in. After some time at the surface, the carbon monoxide level is low—until the animal dives again. In fact, you could probably test the air in the float to get a sense of when the animal has most recently undergone a dive.

The most commonly seen chondrophore is *Velella velella* (figure 3), which has a number of common names, my favorite being "by-the-wind sailor." It gets this name because its float looks much like a small rubbery sail that juts above the raft-like

FIGURE 3. The by-the-wind sailor, *Velella velella*

remainder of the animal. Instead of the larger and singular gas cavity of the man-of-war, the *Velella* float has multiple air-filled chambers and does in fact act like a sail, moving the animal slowly downwind. They do not secrete carbon monoxide into the sail. Instead, the float mostly contains nitrogen. It is thought that both oxygen and nitrogen diffuse through the surface, but that the oxygen is used by the animal for its metabolism, leaving the nitrogen.

The surface float systems of the man-of-war and *Velella* give these animals a number of advantages. First, as mentioned above, they don't have to worry about carefully regulating their buoyancy, since—except for the few times that the man-of-war dives—they are positively buoyant. Second, since wind is common on the open sea, they can travel for free, moving to new locations to satisfy their carnivorous habits. Also, they now live in the two-dimensional surface world rather than the three-dimensional water column. This is a much smaller world and simplifies their finding one another—even if by accident—which is important because men-of-war like to spawn in large aggregations.

There are, however, several downsides. They are vulnerable to predation by sea turtles, which frequent the surface of the ocean. They also are exposed to enormous amounts of ultraviolet radiation. As I mentioned before, open ocean water is freakishly clear, and this extends to the ultraviolet portion of sunlight; however, it is still far better to be 30 feet underwater, using all that liquid above as a sunscreen, than it is to be sitting on the surface like shipwreck victims in an open raft. Men-of-war are known to have compounds that act as biological sunscreens, and *Velella* is likely to as well. Temperature is also a concern, since the top few feet of the ocean can be much warmer than the water below, especially during summer days with low wind. Maybe the biggest problem is the flip side of an advantage—they have no power to move against the wind. The float of the man-of-war is more ball-shaped than that of *Velella,* but also catches the wind. So both animals are pushed willy-nilly about the globe, and can get stranded on land after storms.

Not all of the surface dwellers have obvious floats or any parts that stick above the water. Some have a more subtle approach. There is an amazing purple snail called *Janthina* (figure 4) that hangs from a raft of bubbles glued to its foot, each bubble secreted by the snail and enclosed in a layer of chitin, the same tough material that insects are encased in. Two other animals, as different as can be, both use gastrointestinal gas as their floats. One is *Glaucus,* a tiny silver and purplish-blue slug with what appear to be arms and feet bearing long fingers. Honestly, it may be the most precious animal on earth. It preys on the man-of-war, ingests the stinging cells, and—without setting them off—incorporates them into its skin as a borrowed defense. It appears to hang upside down from the water surface, being kept afloat by air that it swallows into its stomach. The other "gas" floater is special in its own way. It's the manatee. I'm cheating

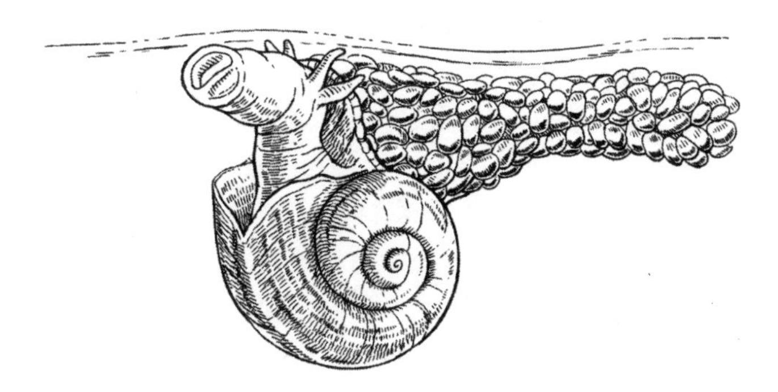

FIGURE 4. A janthinid snail floating under the surface using a bubble raft (*Janthina* sp.)

a bit here, since manatees, and their Indian Ocean relatives the dugongs, live in coastal not oceanic waters, but it's worth it to tell you that they float at the surface using intestinal gas (methane, just as in humans) stored in special cavities. They appear to use flatulence (which releases the air) to dive. I'm cheating even further, a recent scientific publication has documented a previously undescribed beetle from small ponds in New South Wales, Australia, that walks upside down just below the water surface. The flotation appears to be provided by a layer of air on its abdomen, trapped by hairs. This is not unusual in freshwater insects, but what is unusual is that it can get enough purchase from the surface tension of the underside of the water to walk inverted—as if it were the mirror-world version of the water strider we see on ponds.

## The Underwater Balloonists

Of course, the majority of animals in the water column of the open ocean are not bobbing at the surface, but are instead found below it. A number of these, especially fish and siphonophores, use nonrigid gas floats and thus have to deal with the scuba problem of unstable buoyancy. The fish are best understood and often use internal gas floats known as swim bladders. You may have heard of these, and even seen them if you're an angler and have prepared a fish to eat.

Swim bladders, like lungs, originally evolved as out-pockets of the gut. In fact, it has long been thought that lungs and swim bladders are closely related, and that both evolved from the same organ. Strong evidence for this appeared in 2013, when a group of researchers at the University of California at Davis used a specialized version of a medical CT scanner to show that swim bladders in sturgeons, paddlefish, and bowfins received blood from pulmonary arteries, the same way our lungs do. Swim bladders come in two varieties. In about one third of fish species, typically those—like tarpon and herring—that are found closer to the surface of the water, the swim bladder is still connected to the gut by an air tube. This is convenient, because the fish can fill the swim bladder by ingesting air at the surface, and can deflate it for a dive by burping or via flatulence, much like the manatee. So, when you see a fish at the surface gulping air, it may be not because it's oxygen starved but because it has come up from below and is now inflating its swim bladder to remain near the surface. For many fish, though, the surface is either too far away for this to be convenient or inhospitable due to high water temperature, rough seas, or predation. So the remaining two thirds of fish species have a swim bladder that is not connected to the outside world by an air tube. This frees these fish from the need to surface, but introduces new problems. First, the fish needs to be able to inflate the bladder without relying on surface air or, indeed, air in any gaseous form at all. Second, unless the fish accepts being trapped within a narrow depth range, it needs to be able to adjust the amount of air in the bladder as it moves up or down.

We'll address these issues and how the fish solve them in more detail in chapter 3, but I do want to emphasize that this is not a problem just for deep-sea fish. In fact, it is a bigger problem for the animals that we are focusing on in this book—those within 1,000 feet of the surface. The reason for this, as is true for many things, comes down to math. The pressure on our bodies at sea level is about 1 atmosphere, which is approximately 15 pounds per square inch. For every 30 feet you drop below the surface, this pressure goes up by 1 atmosphere. So, suppose you're diving at 30 feet and then surface. At 30 feet, the pressure on you is 2 atmospheres; at the surface it is 1. So, if you didn't exhale during this ascent, the air in your lungs would expand to twice their size. Now suppose instead you are scuba diving at 600 feet (people have done this, though I would recommend against it). To get the same lung-doubling you would have to ascend to roughly 300 feet (20 atmospheres to 10 atmospheres). This same math is why divers need to pay close attention to their ears at the start of a dive, even though this is often the time when they're busy fiddling with their equipment. If they don't pop their ears on the way down to 30 feet, the pressure in the air cavity in their middle ear will be only half the pressure in their outer ear. The eardrum, which is the only barrier between these two parts of the ear, really dislikes this. For this reason, you'll often see divers pause just a few feet down at the start of their dive, trying to pop their ears. Of course, if they can't, they can always surface and wait on the boat, watching the bubbles of everyone else having fun below. The bigger problem occurs if you can't clear your ears on the way up, since you're only a guest underwater and a short-term one at that. So, spending your life at depth isn't an option. This happened to me on a blue-water dive in the Gulf Stream off the shore of Maryland. My ears were able to pop on the way down, but when the dive ended and I ascended, I couldn't pop my left ear at about 30 feet. I kept trying to go a little shallower, but all I got was more pain. Using the universal signal, I pointed to my ear to let my friends know I was stuck. Brad, of flip-flop fame earlier in this chapter, shrugged. There wasn't much to be done. I suppose Brad could have surfaced, gotten a decongestant and a bottle of water (you can, with care, drink underwater and thus take medicine), but I would have been out of air by the time it took effect. So I gritted my teeth and swam upward. It hurt more than I expected, and I expected it to hurt a lot. The pain then stopped, and I felt cold water pour into my middle ear through a new hole in my eardrum. The cold water replicated a procedure known as "caloric stimulation," which led to dramatic effects on my sense of balance. The world felt 100 percent as if it had turned upside down, even though I could see that it was still right side up. My middle ear warmed up this water in a few seconds, and the world switched back. I went the last 10 feet to the surface and was deaf in that ear for some time while the hole healed. Math always wins.

Getting back to the fish, even the ones near the surface need to adjust the amount of air in their bladders as they go up and down, or they will have an uncontrolled ascent or descent, or—worse—they will pop their swim bladder. The bladders are well made (in fact, they were sometimes used as luxury condoms about a hundred years ago), but if the water pressure inside the bladder gets to be over about 2.5 times the pressure at that depth, they start to rupture. More on how the fish manage to inflate and deflate a gas cavity under pressure and without access to gas in chapter 3.

Many siphonophores, as mentioned above, also use gas floats, even if they don't live at the surface. These floats tend to be small relative to body size compared with those in surface-dwelling siphonophores and with the swim bladders of fish, but nonfloating siphonophores are also likely close to neutral buoyancy because they have a higher percentage of water in their tissue compared with fish. In fact, if you dry out a siphonophore, there's not much left. The underwater siphonophores with gas floats that have been studied seem to use the same gas gland process that we discussed for the man-of-war.

I hope I have convinced you by now that nonrigid gas floats have serious limitations, owing to the fact that they compress or expand every time depth is changed. So, what about a rigid gas float? These would not change volume with depth, which solves the instability issue. They turn out to be rare in nature, but there is an outstanding example, the nautilus. The nautilus is a cephalopod—a relative of squid, octopuses, and cuttlefish. It's a distant relative to be sure, with the most recent common ancestor being hundreds of millions of years ago, but it is more closely related to the rest of the cephalopods than to anything else. Sort of like that distant cousin whom you never see but is still family. Unlike other cephalopods, it has a hard shell that at first looks like a snail. If you cut it in half, though, you discover that it's not a snail at all. Inside, you see a perfect spiral lined with mother-of-pearl, divided into compartments, each of which has a small hole that connects it to its neighbors. The animal starts with only about four of these compartments, and continues to add new ones at the large end of the spiral as it grows, living entirely in the outermost and newest compartment.

The nautilus functions almost exactly like a small submarine. Some modern, large submarines have the oomph to power themselves up and down without concerning themselves with buoyancy, but many submarines and possibly all research submersibles have a small number of rigid containers that—when filled with or emptied of air—change their buoyancy and thus the craft's depth. The spirally arranged chambers in the nautilus act the same way. In life, the holes between chambers allow for the passage of a tube known as a siphuncle. This structure controls the amount of fluid in each chamber, thus affecting the buoyancy of the whole animal. It controls the fluid not by moving the fluid itself around, but by moving the salt dissolved within it. Osmosis, which you may have learned about in chemistry class, is a process by which the diffusion of water is affected by the amount of material dissolved in it. I was always terrible at chemistry, so my mnemonic for the process is that "water follows salt," meaning that water molecules will diffuse from a less saline region to a more saline one, assuming they're separated by a barrier that allows the passive passage of water but not salt. This turns out to be useful, because it lets animals move water by first pumping salts across cell membranes. For example, when our digested food reaches the lower intestine, it is still in a fairly liquid form. Our lower intestine then pumps salts and other molecules out of the intestine. This causes the water to follow by osmosis, drying out our stool. Failure of this process leads to diarrhea.

Osmosis, despite requiring only the energy to move the salt, is quite powerful. Using it, the nautilus can control the water level in and thus the buoyancy of each compartment, allowing the animal to go up and down as it pleases. Since the float is rigid, the ascents and descents are more controlled, but the flip side is that the animal's shell can be crushed if it goes too deep. Experiments on these shells in pressure chambers show that adult shells will begin to fail at depths of around 1,000 feet. The shells of very young nautiluses can manage depths up to around

3,000 feet, though. This curious fact, where the small is stronger than the large, is commonly known to engineers who work with pressurized vessels like steam boilers and hot water heaters.

## The Ion Pumpers

When I was young, I was fascinated with both deep space and the deep sea. In other words, I was obsessed with extremes. So the history of the *Trieste* bathyscaphe was one I read over and over, rechecking out the book from the library every two weeks when it was due. In 1960, the *Trieste* reached the bottom of the Challenger Deep, the deepest known spot in the ocean, at about 36,000 feet. This was interesting to me, but more interesting was how it was built. At first glance, the thing made no sense, being a tiny ball hanging below what looked like an enormous submarine. I started reading this book at age seven, so it took me a while to figure it out. The "submarine" was in fact a giant tank filled with gasoline, and the only part of *Trieste* that actually held people was the metal ball, which was barely big enough to stuff two people inside. The ball had thick steel walls, but the "submarine" had walls so thin and flimsy it actually leaked gasoline during storms. I couldn't figure out why two people would need that much gasoline for a craft that only went up and down.

I eventually figured out that I was looking at an underwater blimp, which made sense in retrospect because the family that developed it—the Piccards—were also pioneers in hot air ballooning. The submarine-sized gas tank was not a fuel source but a float—one that would not change size and could not be crushed. The buoyancy provided by it was heavily offset by the weight of the gasoline itself, but because gasoline was lighter than water, a big enough tank of it could provide enough lift to counterbalance the weight of the passenger ball (gasoline was also cheap and readily available on a ship). Since this is not a book about deep-sea exploration, I won't continue with the *Trieste*, but I bring it up because many pelagic animals use the same trick, thus avoiding many of the problems of gas floats.

No animal is known to produce or ingest gasoline or similar organic solvents that are less dense than water, but they do have access to lipids (fats), many of which, as we discussed earlier, are slightly less dense than water—think of your salad dressing with the oil floating on the vinegar. Many eggs of pelagic animals have lipids that serve both as a food source for the developing embryo and also provide buoyancy (more on this in chapter 7). Low-density lipids have been found in pteropods, fish, whales, and many other groups. Certain sharks in particular have unusually large livers. These are up to 30 percent of their body volume and up to 80 percent fat, and thus help compensate for the fact the sharks don't have swim bladders like fish do.

In my opinion, though, the coolest liquid floats are not those made of lipids, but those made of seawater that has been modified by animals to make it lighter. One type of these floats is found in a group of oceanic squid known as cranchiids. These tend to be transparent to the point where you can read a book through them. Like the heteropods, their internal organs are reduced and stuffed into mirrored pouches (more about this in chapter 6). Their eyes are often huge to a comical extent, made even more so by often being on the ends of stalks. Aside from these eyes, the only truly meaty and colored parts of them are their arms and tentacles, which they sometimes raise directly upward over their eyes to better hide them. In brief, cranchiid squid look both stylish and ridiculous. Their transparent bodies consist mostly of a large sac known as a coelom. This sac contains fluid that resembles seawater but on closer inspection is subtly different. We call seawater "salt water" because it contains a high concentration of salt, in particular table salt, which is equal parts sodium and chlorine. In water, the sodium and chlorine separate into sodium and chloride ions. Seawater is about 3.5 percent salt by weight, and sodium and chlorine both weigh more than water, so salt water is about 2.5 percent denser than fresh water. The fluid in the cranchiid coelom, though, is missing much of the sodium that is found in seawater, and in its place is ammonia, which is less dense. Together, this gives the coelomic fluid a density close to that of fresh water, which makes it buoyant in the ocean. The difference is small, so the buoyancy is not large, but it's enough to counterbalance the few parts of the squid that are denser than seawater. One might ask why the squid doesn't simply remove all the ions from the coelomic fluid, turning it into freshwater. The problem is osmosis, which we discussed earlier. If we put freshwater into a sac lined by tissue that has higher salinity, the water will follow the salt that is outside and slowly empty the sac. One could make a sac that is impermeable to water, but it would then be difficult to make the water inside it fresh. The cranchiid squid solves all this by making a fluid that has the same osmolarity (concentration of dissolved material) as the fluid in the rest of its body, but yet is less dense.

Hydromedusae, a large group of animals that are closely related to what we more typically think of as jellyfish, do this as well, as do comb jellies, another gorgeous group of gelatinous pelagic animals, which we will discuss more in chapter 5. These animals remove sulfate ions from seawater. Although the salt in seawater is primarily sodium chloride, other types of salts are in there as well, a major one being magnesium sulfate. The sulfate ion is heavy, being made of an atom of sulfur and four atoms of oxygen, and so would be on the shortlist for removal for any animal that is tinkering with its fluids to become lighter.

As I've said, the buoyancy gained by these changes is not large, and it may be difficult to change rapidly, but it appears to be enough. On blue-water dives, I've seen countless hydromedusae, ctenophores, and other gelatinous animals. Many are so perfectly neutrally buoyant that they appear to have been nailed to the water itself.

## The Endless Swimmers

We end with the animals that need to move to keep from sinking. In my mind, I divide these into the flyers and the frantic strugglers. The fliers are many sharks, fish lacking swim bladders such as mackerel, shrimp that use their antennae like skis, and certain marine mammals. They stay aloft the same way that airplanes do: their forward motion generates enough lift on their fins and other portions of their body to maintain depth. At lower speeds some need to tilt their bodies slightly upward to increase this lift. While it is often assumed that all sharks must move to stay afloat, the sand tiger is a counterexample. This large nurse shark can remain neutrally buoyant while motionless. Like the manatee, it is an air gulper. On one of my favorite dives off the coast of North Carolina, I came upon a sand tiger that was at least 10 feet long. It was about 30 feet below the surface and motionless. Closely surrounding it were nearly a hundred much smaller fish of many types, nearly all motionless as well and facing the same direction. I assume they were scavengers waiting to pick off the remains of anything the sand tiger killed, but for all the world the assemblage looked like a carrier group prepping for a naval battle.

If the sharks and such are the airplanes of the deep, I suppose the frantic strugglers are the helicopters. These are pelagic animals that are negatively buoyant with no means of generating

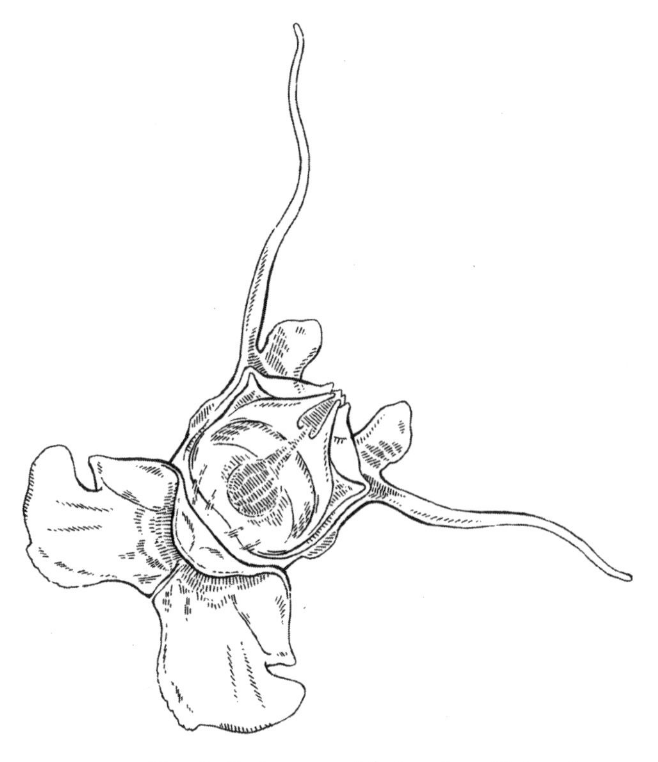

FIGURE 5. The shelled pteropod (winged snail) *Cavolinia*

lift by moving forward. A particularly pitiful example of this is the pteropod *Limacina*. While the pseudothecosome pteropods we discussed earlier have gelatinous shells that likely are close to neutrally buoyant, *Limacina* belongs to the sister group, the thecosome pteropods, which have shells with hard parts. Some of the pteropods in this group—such as the beautiful *Cavolinia* (figure 5)—have shells that are thin, but *Limacina* is swimming with a boulder. Like other pteropods, its snail foot has been modified into two wings, but to my mind these have never seemed adequate to their task. Regardless of my opinion, these snails are always negatively buoyant and must flap hard to stay up. They do get a slight breather every now and then, though. They feed by making a parachute-like mucus web that not only catches things but slows their descent, during which I imagine they are catching their breath. Once they pull in their chute to sort out the food, though, they drop in earnest unless they start flapping. Like Sisyphus, they have a burden that they must continually move up a hill, the only true rest arriving when they die. Then gravity takes its prize, and they fall to the seafloor so far below.

# CHAPTER 3

# Pressure

> Water is not a solid wall, it will not stop you. But water always goes where it wants to go, and nothing in the end can stand against it.
>
> —Margaret Atwood, *The Penelopiad*

## What's Up with That Hole in the Floor?

Scientists are often portrayed as either pedants, force-feeding you things you didn't want to know, or fanatics driven by a singular passion that ends badly for them, if not also for the world. The truth underlying both portrayals is obsession. Some scientists are motivated by ambition, power, or simply the desire to be helpful, but I'd say that the beating heart of most of us is obsession. On my floor at work, I have friends obsessed with ferns, insects, lemurs, mushrooms, and more, devoting their lives to things most people only pay passing attention to. These obsessions often start early in life, and while my love was the ocean, my obsession was machines.

Soon after we came to America, we moved into an apartment building. The apartment was tiny, and the building had collapsing ceilings and was prone to fires, but it had the singular advantage of being on Fifth Avenue, the busiest street in Pittsburgh. The street was only a shadow of its namesake in New York, but

Pittsburgh being the industrial city it was, the road was a parade of heavy equipment ranging from steamrollers to dump trucks to cement mixers. The apartment building didn't have much of a yard, and even that was sometimes covered in ash from the incinerator, so I would sit on the front steps with my mom and count trucks for hours, refusing to go in until I'd seen at least five cement mixers. I've been told that I would get so excited that my hair would grow damp with sweat.

When I went on my first cruise, I immediately became obsessed with the small submersible bolted to the back deck of the ship. Just about every first-timer gets excited about this machine, filling up their cameras and phones with photos of it from every angle while it's being launched and recovered. Indeed, most ships that have a manned submersible treat it like an object of devotion. The research vessel *Atlantis* from Woods Hole takes this to an almost comical extent, being a 274-foot boat with dozens of crewmembers all there to serve the tiny, polished object at the stern. I took this to an extreme, though. When everyone was inside having lunch or sleeping, I'd be walking around the submersible, looking at and touching every part, trying to figure out what it did. It repaid that attention. The *Johnson-Sea-Link* submersible was a wondrous thing, looking for all the world like a helicopter without propellers mated to a bank vault. It had a two-person Plexiglas sphere in the front, where the pilot and a scientist sat, and a two-person aluminum chamber in the back for another scientist and a "rescue" pilot, who could bring the submersible to the surface if the pilot in the front was incapacitated. The two chambers were held together in a cage of aluminum bars but otherwise unconnected, so the submersible was in fact two submersibles forced together like some odd couple. This seemed weird to me. Eventually, one of the sub pilots came out for a smoke one evening, and I started peppering him with questions. Why two subs? Why is one plastic and the other metal? And what's up with the human-sized hole in the bottom of the aluminum chamber? Why didn't people enter through the top, like they did the sphere in front? During the course of his cigarette, he told me the craziest thing.

It turned out that the hole was put in the floor of the rear compartment so it could be opened at depth. They could pressurize the inside of this chamber so that it matched the outside water pressure. Then, when they opened the hatch from the inside, water wouldn't enter. It would just sit there, calm and flat, like a tiny backyard pool. In fact, it was called a "moon pool." Then, and this was the really crazy part, the two people inside would put on scuba gear and slide down through the open hatch and go for a swim outside the submersible, all while they were 1,000 feet below the surface. This was known as "lockout diving." The pressure at this depth is 450 pounds on every square inch of their bodies. My palm is about nine square inches, so pressure at this depth is like holding my SUV in my palm, while my wife's SUV is pushing on the back of my palm just as hard. How does one survive this, and how do animals survive it? To understand this, we first have to think about water.

## Water

Like gravity, water only gets more remarkable the longer you think about it. Whether in a rainbow, mountain fog, rushing stream, or ocean wave, it's stunning in its beauty, especially in its sounds and its interactions with light. When my daughter was young, we would go to the neighborhood pool every day. There was a boy there who didn't interact with others. Instead, he spent each afternoon of the summer splashing water toward the sunlight with a huge smile. I'd often do the same when I was at an outdoor pool as a kid, and always felt that it's only water's common nature that blinds us to what a special substance it is.

Water is also heavy, stunningly so. The water in a full bathtub weighs 350–500 pounds, about as much as a medium upright piano. The water in the Olympic-sized pool I threw up in weighs 5.5 million pounds. This is as much as ten two-story homes. There is nothing in our daily lives that is both as dense and as heavy as water. We deal with things that are heavier, like our cars and refrigerators, but they are not nearly as dense because there's so much air in them. We deal with things that are denser, like kitchen knives and paperweights, but they're typically small. Unless you have to move large boulders, as we do at our farm, or demolish concrete buildings, you'll likely not deal with anything as heavy and dense as the water filling your tub.

We're mostly protected from this knowledge. First, engineers long ago learned to put strong floors under bathrooms and swimming pools that are above the ground floor. If you buy a waterbed, you'll find yourself being probed by the seller about the quality and age of your house, because a queen-sized one weighs 1,500 pounds. Water also tends to be added bit by bit, conveniently through a faucet, so you don't have to start every bath by lugging 500 pounds of it to the bathroom. Finally, water tends to get out of the way. Someone can pour a grand piano's worth of water from the second floor onto you, and you'd be (more or less) okay.

For most, the best indicator of water's true presence is an ocean wave. As I mentioned in chapter 1, I spent part of nearly every childhood summer at the beach, playing in the surf for at least five hours a day. I got good at this and worked my way up to waves that were 10 feet high at times. Every now and then, though, I would make a mistake and get hit by a breaking wave dead-on. Even a 3-foot wave hits hard, and—like everyone who's wiped out in larger surf—bigger waves have held me under the water, face smashed into the sand, waiting for the water to let me breathe again. At sea, our bunks are often right at the water line, and a good wave can slap the hull so hard it makes the whole ship shudder and jolts you out of your sleep.

A classic story of water's true weight is that of the Richelieu Apartments in Pass Christian, Mississippi. This was a large, beachfront, three-story brick building that had been used as a hurricane shelter at times. In general, buildings of this sort hold up well to even strong hurricane winds, but the 24-foot storm surge of Hurricane Camille tore away every single part of this building, leaving only the concrete foundation.

So, water is heavy, both because it is fairly dense and because there is so much of it. For pelagic animals and human explorers, this means that they have to deal with the pressure of it as they drop below the surface. This pressure increases quickly. As we saw in chapter 2, the pressure increases by 1 atmosphere for about every 30 feet of depth. Each atmosphere adds 14.7 more pounds per square inch. Books on the deep sea often focus on how high the pressures get at great depths. They do get astonishingly high: at 8,000 feet it's the weight of a small SUV on every square inch of your body. But what's less appreciated is how high the pressures are at shallower depths. Even at only 600 feet, the pressure is still a hefty 270 pounds per square inch. For someone my weight and height, this adds up to almost 900,000 pounds over my whole body.

Even at shallow scuba depths, pressure can be an issue. Pressure is the reason that snorkels can't be more than a couple of feet long. Longer than that, and your diaphragm doesn't have the strength to inflate your lungs against the pressure of the water against your chest. Going deeper, things get worse quickly. I had a friend who was doing one of her first scuba dives in the ocean.

She was excited, and like some new divers, forgot to breathe as she descended. Normally, scuba divers exhale through their nose, so that the pressure inside the face mask is the same as it is in their lungs and in the surrounding water. However, she hadn't exhaled yet, so the pressure of the air in her face mask was still the same as it was at the surface, but the pressure everywhere around her—and more importantly inside her—was increasing as she dropped. When she got to 20 feet, the pressure inside her mask was 10 pounds per square inch lower than it was everywhere else, and she started to feel her eyeballs being suctioned out of her face. She immediately surfaced. The eye-push had gone away, but the whites of her eyes were now blood red, because the pressure differential had ruptured a number of vessels in the whites of her eyes. Luckily, this—while startling—was ultimately harmless, and she fully recovered, but "mask squeeze," as it's called, can cause more severe injuries. Therefore, even animals and divers relatively close to the surface need to worry about water pressure.

## More Trials of a Scuba Diver

In chapter 2, we discussed the challenges that scuba divers face in controlling their buoyancy. Pressure was always in the background in these discussions, but now we confront it directly.

At first glance, scuba diving seems easy. Once you get good at controlling your buoyancy, gravity goes away, and you're free to float like a balloon over coral reefs, wrecks, and manatee-filled freshwater springs. If the water is warm and clear, and the current mild, diving is magical and effortless. If you want to dive for fun, though, you need extensive training, and if you want to dive for work, you need even more training along with regular physical exams and fitness tests. Although some of it is to deal with situations that could occur in any water sport (cuts, hypothermia, drowning), most of this training is to deal with a threat that is invisible, nonintuitive, and often painless until it becomes lethal or crippling. It is the danger of breathing compressed air.

As we discussed in chapter 2, the only reason we can scuba dive at all is that scuba tanks have a regulator that supplies air that is the exact same pressure as the water around them. This allows us to breathe with the same effort that we breathe on land. If the regulator provided air with a pressure that was low by even a few pounds per square inch, we wouldn't be able to inflate our lungs. If it was low by more than that, the pressure of the surrounding water would start squeezing air from our lungs. If it was really low, the water pressure would crush our rib cage. If the pressure was higher than the water pressure, the opposite would occur, and our lungs would blow up. So, divers appreciate these regulators.

However, this comes with the cost of breathing pressurized air. Recreational scuba divers are often between 60 and 90 feet, which means they are breathing air that has three to four times the pressure of air at sea level. This leads to a number of problems, which can affect both human and animal divers. The first is decompression sickness, often known as "the bends." First seen in free divers and miners a hundred years before scuba was invented, it's perhaps the most dreaded consequence of breathing compressed air, and affects you only when you stop breathing it.

When you breathe gases of any pressure, some gas molecules dissolve in the blood of the capillaries that are found in your lungs. These dissolved gases are transported around the body, where they eventually diffuse out of the blood and into tissues. This is how, for example, oxygen gets from your lungs to your muscles, though your body helps this process along by having a protein in your blood known as hemoglobin that specifically binds oxygen and increases the amount that can be carried. However, 80 percent of the air we breathe is not oxygen but nitrogen. This gas doesn't have a blood protein to bind to, but nevertheless dissolves in your blood, travels around your body, and then diffuses into your tissues, especially fat tissue. So far as we know, this gas is of no use to our body and just sits there. You have about a quart of it in your tissues right now, the level set by the pressure of the air you breathe via a chemistry principle known as Henry's law. Most people will never be aware of this, and their personal quart of nitrogen will be with them until they die (and beyond, in some cases). If you were to suddenly travel to altitudes higher than 18,000 feet, though, and especially higher than 25,000 feet, you might be in trouble. As I mentioned, the amount of gas that can be dissolved in blood or tissue depends on the pressure of the air you breathe. If that pressure drops rapidly, gases start coming out of solution in the form of bubbles, which is why the bends is often likened to the opening of a carbonated beverage.

It is impossible to predict fully where the bubbles will form. In some locations, the bubbles do not cause permanent harm. Those that form in the skin usually cause itching; those that form in the joints cause pain, which can be severe, but typically don't cause lasting damage. Bubbles in the veins can cause lung pain and damage when they eventually get there, but the worst effects are reserved for bubbles that appear in the brain or spine. These can lead to seizures, strokes, and permanent neurological damage (partial or complete paralysis, incontinence, hearing loss).

In the early days of airplane and balloon travel, the pilots were vulnerable to the bends. Cockpits were not always pressurized, and early balloonists often rose like Icarus to great altitudes. These days it is unlikely that you will ever ascend rapidly to a height that will cause this problem. We do indeed spend our lives "diving" at the bottom of a 100-mile-deep ocean of air, but luckily we seldom ascend rapidly enough to realize this. However, although we would have to ascend from sea level to 17,000 feet (the altitude of Everest Base Camp) for the pressure to be half of what it was, a scuba diver does this by simply ascending from about 30 feet to the surface. A diver returning from 100 feet to the surface sees a fourfold drop in pressure, which would be the same as a sea-level person suddenly ascending to 34,000 feet, which is 5000 feet higher than Everest. So the bends is an omnipresent problem in scuba diving. Even those who dive only to shallow depths can be affected if they stay down too long, and even free divers who don't carry scuba tanks can run into trouble if they dive over and over in the same day, because the air in their lungs will be compressed at depth each time they go down. In fact, the earliest cases of the bends were likely among pearl divers, who repeatedly free dove to collect oysters at relatively shallow depths. Even the bones of deep-diving whales can be riddled with cavities thought to be due to expanding gas.

Thus, much of scuba training involves minimizing the risk of the bends, by controlling both your depth, how long you are at depth, and especially how quickly you return to the surface. The deeper you go, the less time you can spend at depth. Like the rabbit in *Alice in Wonderland*, we're forever looking at our watches when we're underwater. If you go beyond the allotted time (which is set by various tables), you have to stop at a sequence of depths on the way up to allow the extra nitrogen in your blood to come out of solution slowly enough that it doesn't form bubbles. Going back to the carbonated beverage, it's like opening the bottle slowly so that it doesn't fizz uncontrollably.

Nitrogen has a second danger, which makes maintaining this control over depth and ascent more difficult. This is called nitrogen narcosis, and is essentially getting drunk underwater. It turns out that all gases, except for helium and possibly neon, act like alcohol when breathed at high pressure. How strongly they do this depends a lot on how well the gas dissolves in fat, and some are far more potent as intoxicants than nitrogen. Xenon gas in particular is used as a surgical anesthetic in certain cases, because less than 1 atmosphere of pressure of it will put someone out. So, as is the case with the bends, nitrogen isn't special; there just happens to be a lot of it, so pressure due to it gets high more quickly with depth.

There is a rule of thumb, known as "Martini's law," that says that the intoxicating effect of nitrogen is equal to one standard drink for every 30 feet below 60 feet. In truth, people vary in their vulnerability to nitrogen narcosis, but typically you start to see significant effects once you're below 100 feet. I seldom dive to this depth and have only been mildly "narced." To me, it does feel like being drunk, but with more euphoria and less lethargy. I dove once with a graduate student and watched her get narced at 110 feet. I looked over and she was just floating in place with a blank stare. I couldn't get a response from her, so I grabbed her by her buoyancy control vest and pulled her up to 90 feet, where she recovered.

Nitrogen narcosis can also happen in hyperbaric chambers. When I first came to graduate school, I went down to a hyperbaric chamber facility in Wilmington with an older graduate student and my advisor, Bill. The student was interested in how certain snails known as limpets attached themselves to rocks and theorized that they were using suction, which would mean that they would suck harder at depth. Rather than doing this work underwater with scuba equipment, which would have been a mess, he decided to use a dive chamber. Because a pressure chamber has high levels of oxygen, fire safety is a concern, which meant they couldn't have electronic recording equipment inside the chamber. They needed me to stand outside the window of the chamber and write down the numbers they told me. All went well, until we started doing some deep "dives." My normally sober advisor and lab mate then started giggling hysterically. This got worse, and eventually they were writing funny notes and slapping them against the window of the chamber. As the chamber pressure dropped back down to 1 atmosphere, they went back to their former and more formal selves. They wouldn't even admit to having been narced, but I know what I saw.

As you can see, nitrogen narcosis feels good; and in a dive chamber it is safe. On a scuba dive, though, especially a blue-water dive over deep water, it can be lethally dangerous. Like all forms of intoxication it impairs judgment, and in this case is doing so in an alien situation where judgment is the only thing between you and death. Even worse, it seems to invoke a specific desire to become one with the ocean and to go ever deeper. This is why nitrogen narcosis is sometimes called "the rapture of the deep." Even the mild narcosis I have felt gave me the feeling that I was a magic fish, invulnerable to harm. At deeper depths, divers have been known to take out their breathing regulators, and in blue-water diving situations there is a serious risk of the person unclipping from their tether and simply swimming down until they die. As with alcohol, many people who are narced will not admit they are, so, like the bends, nitrogen narcosis is an insidious danger and another reason that divers are so carefully trained.

Then why don't we use 100 percent oxygen to dive? The problem is that oxygen is also dangerous at high pressure.

Despite our absolute need for it, too much oxygen is toxic. Much of this toxicity is related to the fundamental use of oxygen in our cells, which is to oxidize other molecules. In a very real way, to live is to slowly burn, and our cells—using mitochondria, which they made a symbiotic bargain with billions of years ago—are exquisitely good at controlling these slow oxidative processes. However, too much oxygen creates a situation where molecules are oxidized in an uncontrolled fashion. This can happen even at normal air pressures, and any trip to a health-food store will put you in front of a shelf with antioxidant foods that are advertised to control oxidative damage. A diver at depth is breathing oxygen at a high pressure. If one goes deep enough, usually well below 100 feet, this oxygen can cause problems. The primary sites of damage are the lungs, the nervous system, and the retinas. In the lungs, oxygen toxicity leads to difficult and painful breathing. At high enough pressures, the alveoli (the small sacs in our lungs where gas exchange occurs) start to collapse, leading to suffocation. In the eyes, oxygen toxicity leads to vision changes and retinal detachment. In the rest of the nervous system, it can lead to seizures. Given that diving requires continual conscious and sober control of a multitude of factors, seizures at depth are extraordinarily dangerous.

So, nitrogen and oxygen, while our friends (or at least innocuous bystanders) at surface pressures, can be dangerous at depth, in ways that are not only nonintuitive but also creep up on us without warning. So, how did the divers that swam out through that hole in the floor of the *Johnson-Sea-Link* submersible at depths of many hundreds of feet do it? They should have suffered from terrible oxygen toxicity and enough nitrogen narcosis to knock them unconscious. They escaped these dangers by breathing a special mix of gas that is mostly helium, with the amount of oxygen and nitrogen kept low enough that when they breathed the total gas at high pressure they didn't get the ill effects of either gas. Helium can give you the bends even more so than nitrogen, though. So when the divers returned to the dive chamber, the chamber was sealed and kept at the same high pressure all the way to the surface. After the submersible was recovered by the ship, it was lowered on top of a small decompression chamber that docked with the hatch of the submersible. This hatch let the divers move into the decompression chamber. They spent the next twenty-four hours trapped in a double-wide metal coffin on a deck that was sitting out in the tropical sun as engineers slowly lowered the pressure inside the chamber back to 1 atmosphere. I'm not claustrophobic, but I can't think of a worse way to spend a day.

I've gone into the challenges of scuba diving not to scare you away from the sport, but to impress on you that pressure underwater is a real concern, even at depths of only 100 feet. Yet pelagic animals seem to manage just fine. Fish are known to live at depths up to 26,000 feet, and many marine mammals (and even some birds) dive to great depths over and over. They do this using a suite of adaptations, some of which are disarmingly simple. We first deal with the animals that don't contain air in gas form, and then those that do.

## *Nothing but Water*

In chapter 2, I suggested that water is incompressible. This isn't true. Water *is* in fact compressible. The compressibility of a substance is usually quantified by a value called the "bulk modulus." The bulk modulus of water is 300,000 pounds per square inch. This is indeed a high number and means that even 2 miles underwater, which is the average depth of the ocean, the water is compressed by only 2 percent. But water is not unusually hard to compress as liquids and solids go. Its bulk modulus is about the same as that of other fluids, and is much less than that of many solids. For example, the bulk modulus of certain human bones is ten times higher, and that of common steel is a hundred times higher. Diamond has perhaps the highest bulk modulus, at over two hundred times that of water. But we don't have an ocean made of diamond, steel, or even bone. Going back to our discussion of water's weight, water is not unusually incompressible for a liquid or solid, but it is the most incompressible substance you could imagine having enough of to fill an ocean with. Water is also far less compressible than air, by a factor of about twenty thousand.

I travel a lot and so have sat next to countless taxi drivers and passengers in planes, trains, and automobiles. When I tell them that I'm a marine biologist, they inevitably ask me one of two questions: (1) "Have you ever seen a mermaid?" and (2) "How do you keep the deep-sea animals from exploding when they come up?" I've never known what to do with the first question—sometimes I mumble something about manatees. The answer to the second question is easy: we do nothing. If I caught a sea urchin from 3,000 feet down and handed it to you, you could put it in your home saltwater aquarium and feed it carrots. As long as you kept the water cold, it'd live in your aquarium for months, maybe longer if you were careful. This isn't always true for animals that live deeper than 6,000 feet or so. At these depths, the pressure is high enough that protein structure is subtly different, and cell membranes have different composition. These deeper animals therefore might not survive a trip to sea-level pressure. I mostly work with animals that live in the top couple of thousand feet, where we don't worry about pressure. This is possible for two reasons.

First, as those ads that tell you to drink eight cups a day like to remind you, the primary component of most animals is water. The brain and the (nonair) parts of the lungs are about 75 percent water, muscles are 80 percent water, and skin is 64 percent. Even bones are about 30 percent water. Just as important, the remaining components of these tissues are at least as incompressible as water. So, for those tissues that don't contain gas, water pressures down to impressive depths don't cause trouble. This is why the lockout divers going for a swim at 1,000 feet weren't in any danger of being crushed as long as their lungs (and middle ears) contained air of the right pressure.

Second, and this is far less appreciated but just as important, the pressure of the oxygen dissolved in water is not affected by water pressure. This is very different from the situation on land, where the pressure of oxygen depends strongly on your altitude. As I said before, we live our lives at the bottom of a 100-mile-deep sea of air. But this is a squishy sea, where the air at the bottom is compressed owing to the weight of all the air above it. The oxygen dissolved in ocean water does not have a pressure as such. Instead, it has something called tension. Tension has a specialized definition that we won't go into, but the important facts for our purposes are: (1) it is related to the amount of gas dissolved in the water, and (2) it is not affected by the water pressure. Gas tension can change with depth for a number of reasons, including currents and biological factors, but it does not keep going up with depth. Therefore, if you get your air in a nongaseous form (i.e., without lungs), your blood and tissues don't have the large concentrations that lead to the bends, nitrogen narcosis, and oxygen toxicity.

This means that pelagic animals without air cavities don't have to worry about pressure at all, at least if they are living in the top 6,000 feet or so. You may think that the buildup

I presented of the dangers of water pressure is like a shaggy-dog story—overly long and with a bad punch line. I find it fascinating, though, that the very first thing you'd think would be a problem isn't a problem at all. The majority of animals in the ocean don't have any air cavities to crush, and also use gills (or their skin) to absorb oxygen. So they're protected from pressure. What about the animals that do need lungs to breathe, or those that use air floats, such as those we discussed in chapter 2? It turns out that they manage via a number of adaptations, some complex and nonintuitive, and some surprisingly simple.

## Pressure and Air-Breathing Animals

There are of course no animals breathing compressed air from scuba tanks—aside from the occasional freakishly loyal and well-trained dog—but there are many air-breathing diving animals. Aquatic adaptations are found in all mammals except rabbits, and certain primates, but diving mostly evolved in only three groups; manatees/dugongs (related to elephants), seals/otters (related to weasels, ferrets, and wolverines), and whales/dolphins (related to hippos). The manatees and dugongs do not appear to dive deeply, but some of the seals and whales dive to extraordinary depths. Most can dive to well over 1,000 feet, but the real champions are sperm whales, beaked whales, and elephant seals. The sperm whale, the largest carnivore on the planet by a hefty margin, has been shown to dive to about 7,000 feet, which is the same depth that elephant seals can reach. A number of beaked whale species are in the same range, but the current depth record holder for a diving air-breather is Cuvier's beaked whale, which can get down to 9,800 feet.

As the old saying goes, "Believe half of what you see and none of what you hear." To my knowledge, no one has seen a

Cuvier's beaked whale at 9,800 feet. Only a few submersibles go down to this depth, and these whales are rare enough that the chances of someone looking out of the porthole of one of these craft and seeing a whale go by are vanishingly small. Sonar might give you a clue that something is down there, but not a good idea of what it is. So how do we know? It turns out that certain labs, such as Peter Tyack's at Woods Hole Oceanographic Institution and Andy Read's at the Duke Marine Lab, have developed tags that, once attached to the whale, record depth, speed, and a number of other factors as they dive. Of course, the next question is, how do you get the tag on the whale? Tagging large land animals is difficult enough, and if you're my age you remember the afternoon animal shows where they had to tranquilize the animals first. This might work with an elephant seal, but obviously won't work with a whale. In 2014, I did a sabbatical in Peter Tyack's lab while he was at the University of St. Andrews, and he explained it to me. Essentially you find a whale (or group of whales) relaxing at the surface. You then launch a small boat, preferably a rowboat, and sneak up on the whale(s). You can get only so close, so you use a very long pole, a bow and arrow, or a low-power pneumatic rifle to get the tag on the whale, where it hopefully will remain attached when the animal dives. Many tags are held on only with suction cups, so they don't stay on long, maybe only for one dive, but some can stay on long enough to give us a sense of the animal's diving behavior. As you might guess, the failure rate at all stages of this process is high. In classic scientific dispassion, many scientific articles where these methods are used simply say, "whales were tagged." It's worth knowing how hard these things are to do.

Mammals are not the only air-breathing divers in the ocean. Quite a few birds dive to moderate depths. One of my favorite photos is of a small school of fish being attacked by sharks from below and gannets from above. The champion divers in the bird world are the penguins, especially the emperor penguins. These flightless birds are 3 feet tall—the average height of a two-year-old—and can weigh 100 pounds. They've also been shown to dive to depths of at least 1,800 feet. Like most scientists, my knowledge is specialized, and I've never been an expert on birds. So when I learned how deeply these animals could dive, all I could say was, "Really?" I had assumed that all penguins dove to maybe a few dozen feet. It was like knowing that cats were good jumpers, but then learning that they could clear the Empire State Building.

So, there are air-breathing animals that can dive to extraordinary depths. Even the whales, dolphins, penguins, and seals that dive to depths of only a few hundred feet are well past the danger limit for oxygen toxicity, nitrogen narcosis, and the bends. You might guess that these animals are immune to these particular dangers, but they're not. They also, like human free divers, are vulnerable to these dangers because the water pressure will pressurize any air in their lungs once they reach depth.

The first step to understanding what is going on here is knowing that the lungs of diving mammals are actually *smaller* than those of similarly sized land animals. These animals are going to extreme depths on a single breath of air, sometimes for hours at a time, so you'd assume that their lungs would be gigantic, but they're in fact unusually small for the size of the animal. Their lungs are also far more muscular than those of land animals, allowing them to exhale up to 90 percent of the air in their lungs. Humans, by comparison, exhale only about 10–15 percent of their air.

The stunning fact is that most marine mammals exhale before beginning a deep dive. They also exhale so forcefully that there is almost no air left in the lungs before they go underwater.

Water pressure takes over and completely collapses the lungs once the animals have reached a depth of about 100 feet. Instead of following the complex procedures and limits that human scuba divers undergo to protect themselves from the bends, narcosis, and oxygen toxicity, they simply opt out of the game altogether by getting rid of all their air. It's the ultimate "take my ball and go home" strategy, and solves all the problems of breathing compressed air by simply not doing it. Because of this, they can dive to depths well beyond the limit of any scuba diver, even one using exotic gas mixtures.

Of course, there is one nagging problem: where are they getting the oxygen to live? By solving the compressed air problem, the deep diving mammals have created an even bigger one—suffocation, made worse by the fact their metabolisms can be up to three times higher than that of land mammals of the same size. Their solutions to this issue are manifold and vary somewhat by species, so I'm only going to hit the highlights. First, despite being unable to store oxygen in their lungs, these animals nevertheless store a great deal of oxygen in their bodies. The oxygen is stored in their blood, both bound to hemoglobin and dissolved as gas, and also in their muscle tissue, bound to myoglobin. Myoglobin, which is what makes red meat red, is one-fourth of a hemoglobin molecule, or I suppose you could say that one hemoglobin is four myoglobins stuck together. However you think about it, myoglobin—like hemoglobin—binds oxygen, acting like a local reservoir for muscle cells. Of course, humans store oxygen in blood and muscle as well, but the relative amounts are different. A human may have 25 percent of their oxygen in their lungs, 60 percent in their blood, and 15 percent in their muscle, while an elephant seal has only 5 percent in its lungs, 70 percent in its blood, and 25 percent in its muscle. Some diving mammals have 40 percent of their oxygen bound to myoglobin in their muscle. They do this using such high concentrations of myoglobin that their meat looks dark brown. Others have circulatory systems with a lot more blood or more hemoglobin per blood volume than found in similarly sized land animals. Some do both tricks, but in general, diving mammals have ways to store oxygen that are well above what we can do. Many seals, for example, have twice the amount of oxygen in their bodies than we do, even with their deflated lungs.

Second, if they're forced to stay down for a long time, some diving mammals have an impressive ability to maintain their cellular activity without oxygen. This is known as anaerobic metabolism and is found in many organisms; diving mammals are simply better at it. It is much less efficient than aerobic metabolism and also comes at a cost, which is lactic acid buildup in tissues. This happens to human athletes as well when they push themselves beyond the capacity of their oxygen supply and leads to leg pain and cramps. Marine mammals can let their lactic acid levels get very high by human standards, but at the end of a long dive of this sort, the animals need time to recover. Interestingly, though, some marine mammals are able to dive to great depths over and over for relatively long periods without requiring any rest. This leads us to a third adaptation, known as the dive reflex.

There's a ghoulish saying among some rescue workers that one is "never cold and dead, only warm and dead." This comes from the fact that occasionally a person who has been recovered—apparently dead—from cold water later seems to come back to life as they warm up. This is known as the mammalian dive reflex, and it used to be a lab exercise in my undergraduate physiology class. One of the two lab partners would stick their face into a bucket of ice water while the other measured their pulse. We had to stop the exercise after a couple of years because some of the male students turned it into a contest to see who could last in the bucket the longest, but when it worked you would see a sharp drop in heart rate. What we weren't able to see was that blood was being redirected all over the body via changes in blood vessel size.

Our blood vessels control where blood goes by making themselves narrower or wider, known respectively as vasoconstriction and vasodilation. You see vasoconstriction when your face goes white with fear, and vasodilation when your skin gets warm and red when exercising on a hot day. In the dive reflex, the blood vessels to all the less important parts of our bodies (e.g., our muscles and digestive systems) are constricted, and the blood vessels to the parts that matter (e.g., the brain) open up. At the same time, heart rate and body temperature drop to the point where someone might think you're dead. This effect is strongest in younger people, especially small children. In adults over about 25, it's harder to see, but it is well developed in diving mammals, allowing them to extend their dive time by sending blood and oxygen only to the tissues that need it, which in their case is primarily the brain and sense organs. The swimming muscles surprisingly do not receive much blood, and instead get their oxygen from the myoglobin in those tissues.

Less is known about how diving birds manage this feat, especially the emperor penguin, which can spend half an hour at great depths. It appears that it has special biochemical adaptations in its hemoglobin that allows it to supply its tissues with oxygen, even when there is little oxygen to be found. Diving mammals have similar biochemical adaptations that allow them to push the limits on their dives.

Overall, diving air-breathing animals have the exact opposite problems of human scuba divers. We have plenty of air (as long as it lasts) but suffer from the challenges of breathing it at the pressure required to prevent our lungs from being crushed. In contrast, diving animals dodge the pressure issues but at the cost of an impoverished air supply. When it comes to dive performance, however, there is no comparison. Even with thousands of dollars of specialized gear, we can't hope to come even close to matching what these natural divers accomplish.

## Blowing Up a Balloon while Standing on It

The last thing I want to discuss is the swim bladder of fish. As we mentioned in chapter 2, the swim bladders of most fish are not connected to the outside world and must be inflated in a different way. For many of these fish, it means they must be inflated against some serious pressure. They do this using the same principle found in nuclear reactors.

Nuclear reactors are (at heart, at least) simple contraptions. A nuclear reaction generates heat. This heat is used to boil water, which creates steam. This steam has pressure, which is used to spin a fanlike turbine, which is connected to a generator, which makes electricity. It's not so different from using the steam of a boiling tea kettle to spin a small fan, except you replace the stove with a slow-burning nuclear bomb. The problem—one of many, actually—is that the water that is heated by the nuclear reaction becomes radioactive. You don't want steam from this water flowing through all your nice electric generator parts. This means you need to transfer the heat from the radioactive water to clean water without the two waters touching. Your first thought might be to put the two waters in two pipes and have them side by side, so the radioactive water heats up the clean water as it flows by. But which way should the waters in the two pipes go? They could both go in the same direction or in opposite directions to each other. If the waters both flow in the same direction, let's say left to right, then at the left side you have hot radioactive water and cold clean water. Heat goes from hot to cold, and so as we move right, the radioactive water gets colder and the clean water gets hotter. However, at some distance from the left, the waters will be at the same temperature and no more heat will flow. This works, but not as well as we'd like. What if the waters went in the opposite direction? Then, at the left, the radioactive water is hot and slowly cools as it moves right. The clean water at the right is the coldest, and so even though the radioactive water has cooled, it's still warmer than the clean water and so dumps some heat to it. The clean water on the left is at its warmest, but still cooler than the radioactive water, which is also at its hottest, so again heat is transferred. Overall, a lot more heat gets moved from the one pipe to the other.

This is known as a countercurrent exchanger and shows up everywhere in biology and technology. For example, have you ever wondered about those wading winter birds with their feet in freezing water? How do they keep from losing all their heat through their submerged feet? It turns out that there are countercurrent heat exchangers in their legs in the form of neighboring arteries and veins. So as warm blood is leaving the core of the body in an artery, the heat is efficiently transferred to the blood coming back up the legs into the body core through a vein. This leaves the feet cold indeed, and you have to feel bad for the bird, but it keeps the body from going hypothermic.

The mother of all countercurrent exchangers is found in the walls of the swim bladder of fish. It's called the "rete mirabile," which means "miraculous net," and it's miraculous indeed. The one in the common eel is about the size of a raindrop, but contains a countercurrent exchange system of two hundred thousand arterial and venous capillaries, which—if put together end to end—would be half a mile long. The rete mirabile serves two primary functions: (1) to keep the gas inside the swim bladder, and (2) to inflate the swim bladder.

Let's start with keeping the gas in the swim bladder. The air pressure in a swim bladder in a fish that is 600 feet underwater is about 21 atmospheres. This gas is next to the venous capillaries of the rete mirabile, and because of its enormous pressure, much of it is dissolved into the blood. This venous blood is now leaving the swim bladder; if this process were to continue the bladder would soon lose all its gas. However, the supercharged blood must first pass by the countercurrent exchange system of the rete mirabile. As with the heat examples in the bird and the nuclear reactor, the exchange is so efficient that pretty much all the dissolved gas is transferred to the arterial capillaries and sent right back into the swim bladder. In a way, the rete mirabile acts like a revolving door; just as you think you're leaving the building, you stay in the door too long and are rotated back inside. It also reminds me of bridges in Pittsburgh, whose off-ramps often immediately lead to an on-ramp on the same bridge, sending you back the way you came.

The question, of course, is, why have blood vessels next to the swim bladder in the first place? The rest of the swim bladder is covered in guanine crystals. This is the same material that makes silvery fish shine and, in this case, makes the swim bladder impermeable to oxygen—except at the location of the rete mirabile. The rete sits there like an open drain in a bathtub, but needs to be there because this is how the bladder is filled in the first place. This is again done using the rete, but with a twist. The twist involves the definition of tension, which I did not give you earlier. I said that tension was related to the quantity of gas dissolved in a solution, but it's more accurate to say that tension describes how strongly a gas will move from one solution to another or from the solution back into the air. In certain odd situations, one solution can have a lower concentration of gas than another, but nevertheless a higher tension. In these situations, the gas will move from the low concentration solution to the higher one, which is a bit like your bike coasting uphill.

What happens in the swim bladder is this. The arterial capillaries are bringing in oxygen, and the venous capillaries are taking oxygen out. Much of this oxygen is bound to hemoglobin, and because of the countercurrent system in the rete, the oxygen doesn't go back to the fish's body. This is good, but it won't inflate the bladder. When the fish wants to inflate the bladder further, it secretes lactic acid into the venous capillaries, making that blood acidic. When this happens, the hemoglobin loses its love for oxygen and lets it go. This process, known as the Root effect, is extremely strong in fish. This raises the tension of the oxygen in the venous capillaries, which sends oxygen over to their next-door neighbors the arterial capillaries, even though the arteries have more of it. The fish also secretes salt into the veins, which forces the nitrogen out of them into the arteries. The countercurrent structure of the system makes both of these processes very powerful, and the whole system is known as a countercurrent multiplier. In the common eel, which has been studied in detail, the process is so strong that it can inflate a swim bladder against 3,000 atmospheres of pressure, which is far more pressure than is found in even the deepest parts of our ocean.

The rete mirabile lives up to its miraculous name, but can do its job only slowly, over hours or even days. This is why fish that are caught in nets often have swim bladders that have become too big for the body or that even rupture. As amazing as the adaptations of many animals are, they usually work only within the bounds of what the animals typically experience in their lives, and no fish evolved to survive being suddenly swallowed by a net and pulled thousands of meters to the surface. This is why biologists often say that life is both robust and delicate.

# CHAPTER 4

# Light

However vast the darkness, we must supply our own light.

—Stanley Kubrick, interview in *Playboy*, 1968

## My First Dive in a Submersible

I didn't get to dive in a submersible during my first research cruise and had to instead live vicariously through the dives of others. My first dive was a few months later in the Bahamas with a different lab led by Craig Young, who studied the puzzle of how deep-sea animals knew it was spring and thus time to breed. He kindly offered me a spot on the ship, where an even kinder graduate student offered me one of his spots on a dive, which was in the glorious front chamber. After the cruise, I was so grateful that I gave him one of my two guitars, which I think made him uncomfortable.

Submersibles are the true lead zeppelins: large, heavy, and cumbersome on deck and in the air, but as peaceful as a hot-air balloon under the water. After a morning of avoiding coffee (a dive is four hours long), I traded my usual T-shirt and shorts for jeans and a sweatshirt to deal with the cold at depth and went to the galley where the submersible's crew held their usual predive meeting. This mostly consisted of reminding everyone of how deep we were going, how long we'd be down, what we planned to do, and—importantly—how much each of us weighed that morning. Neutral buoyancy is as important to submersibles as it is to scuba divers, and a few pounds error can be the difference between a smooth dive and spending four hours at an angle.

After this, the pilot and I climbed into the front chamber of the *Johnson-Sea-Link* through a hatch in the top, while a second scientist and a backup pilot climbed into the rear chamber through the moon pool hatch in the bottom. We tried to do this quickly, because it was already hot and even more humid, and we were now dressed for a crisp fall day. Both hatches were then sealed, the air-conditioning was thankfully turned on, various communication and safety checks were performed, and the submersible was lifted by a heavy-duty winch mechanism and slowly lowered into the water. Once it hit the water, the first thing I noticed was that we were looking right at the ship's large propellers, which seemed too close and were still spinning. We were also wobbling around in the waves in a nauseating way. The tether was released, the ballast tanks filled with water, and we started to go down.

Some moments you always remember. I remember the first time I kissed my wife, Lynn, sitting side by side in the back of her station wagon after I nearly lost her in a dark North Carolina swamp in March. Seven years later, almost to the day, I remember my daughter Zoë being cut out of Lynn in an emergency caesarian delivery, while Lynn—on an epidural and with a sheet blocking her ability to look down—smiled up at me. In between those two life-altering moments, I remember every moment of that first submersible dive. I remember the quiet, the blinking of dozens of red panel lights, and the sub pilot taking my photo and giving me a cherry-flavored candy. But what I remember most is the light. This was before I had done a blue-water scuba dive, so I had never seen what the open ocean looked like underwater. I had dived on Florida coral reefs and spent a teen summer with my Bosnian relatives swimming in the clear Adriatic off Croatia, but I'd never actually seen the true open ocean. This cruise and the previous one were the only times I had ever even been on the ocean since my passage to America as a baby. I spent a lot of time looking over the side during both research cruises, mostly because I was so shocked by the water color. Nothing prepares you for it. This water, only a few feet away from me standing on the deck of the ship, was like an undulating blue jewel. Sapphire earth.

My first submersible dive took me inside this jewel, one that was lit by the sun in all directions. I spent nearly the entire dive down looking in all directions to understand what the light was doing. For the first 20 feet or so of depth, it was very much like being in a clear swimming pool, with lots of lighter and darker stripes racing across the submersible. These stripes, known as caustics, are caused by the waves acting as lenses for the sunlight. When I looked straight ahead, the water was blue, and when I looked down it was violet with slowly waving rays of slightly brighter light seeming to come from a point directly below me. Looking directly up was impossible, like staring into a floodlight, but if I looked up at an angle I saw the mirrored underside of the water undulating with the waves. These waves still pushed us around, and the pilot was having many quick chats with the ship's bridge during the start of the dive, so the peace I mentioned hadn't started yet.

Once we dropped to about 100 feet, things changed. Looking forward and down, the light was still blue and violet, but it was possible now to look straight up. I could no longer see the waves or the caustics or the rays coming up from below. Everything was smooth and still, with the only feature outside the submersible being the slow gradation of light from deep violet to cobalt blue to sky blue as one looked from straight down to straight up. Also, and this took a moment to realize, all the color was gone from inside the front chamber of the submersible. Aside from the red LED indicator lights, everything was just a lighter or darker shade of blue. You might think that by this point the light would have been dim, but it wasn't. It wasn't as bright as being on the deck of the ship, but it was brighter than being in a room. It was just very blue. I was so astonished by it all that I dropped my half-finished candy, which promptly rolled between our two seats and down into the bowels of the instrumentation at the bottom of the front chamber.

For the next few hundred feet, the light slowly got dimmer. It also got even bluer, which didn't seem possible. Even at 600 feet down, it was bright enough to read a book by. By about 1,000 feet, the blueness began to fade into gray because it was getting dark. It wasn't truly night yet, more like late twilight. First the view below went from purplish blue to very dark gray to black. Then the view ahead did the same thing, leaving only light from above, which also slowly turned gray. Eventually, somewhere around 2,000 feet, the only way I could tell there was any light at all was because I could see the silhouette of the aluminum cage that cradled the front chamber.

Long before the sunlight went away, though, another form of light showed up, which was bioluminescence from the plankton we were dropping through. I was used to bioluminescence from fireflies, which is bright, greenish-yellow, and slowly blinks on and off in our twilight pastures. But these were thin streaks of a blue that was just as pure as the background light, only brighter. They were highly varied in shape; some just looking like little blinks while others looked like small fireworks. Some looked like two tiny carnival rides had smashed into each other, with incandescent girders and wheels flying off in all directions. All this light was only next to the submersible itself. The plankton apparently had to hit the various parts of our vehicle before they would glow. The water beyond the submersible remained black.

The submersible pilot turned on the lights, landed on the sea-floor, and did his job, which was far less interesting to me than the ride down. After about three hours, the submersible returned to the surface and the world went from black punctuated by blue, spaghetti-like flashes of bioluminescence, to gray, to blue, and back to the intense brightness and warmth of the tropical day. Again, we were staring at the all-too-close rotating propellers and bobbing in a nauseating way, but this time one of the sub crew dove into the water to reattach the thick rope to the top of the submersible. The submersible was pulled out of the water and set back on the deck, where I saw one of the crew approach us with a long grabby thing that was needed to extract my sticky candy from among the instrumentation at the bottom of the chamber. They still remind me of that.

## Happy Families

I got tonsillitis often as a child, which would give me high enough fevers to hallucinate dinosaur cartoons on the wall next to my bed. This kept me home from school, but my mom wasn't a fan of my lying idle, so she'd give me book assignments. My first was *To Kill a Mockingbird,* which was far over my seven-year-old head. Another was *Star Wars,* which as you can see by now affects my imagination to this day. A third was *Anna Karenina,* which opens with, "Happy families are all alike; every unhappy family is unhappy in its own way." Marine light is like that. All ocean waters look the same, but each coastal water is different in its own way.

What water looks like, both from the surface and from inside (when we are submerged), depends on the optical properties of water itself and the stuff that happens to be in it. In the open ocean, especially in the large tropical regions, the water is mostly just that—water. As I've mentioned before, water is miraculous in a number of ways. Optically, it's miraculous because although we think of it as so clear, it's actually opaque to just about all radiation. Electromagnetic radiation is a large family of energy-transmitting waves, some of which are miles long, some of which are shorter than atomic nuclei are wide. The long waves we call radio; some of the very longest of those are used to communicate with submarines and require transmitters that are up to 40 miles long. From there, the waves steadily get shorter as we go from AM radio to FM radio, to broadcast television, to cell phones, radar, the microwaves that heat up your leftovers, and infrared waves that you're emitting right now as a warm-blooded mammal. At the short end, we have ultraviolet waves, which burn our skin if we stay at the beach too long, X-rays at the hospital, and finally deadly gamma rays from nuclear reactors and atomic bombs, whose waves are smaller than atoms. From the miles-long radio waves to the subatomic gamma waves, there is only a tiny region of the electromagnetic spectrum to which water is fairly transparent. We call this the visible region, and for good reason. Because we are also made of water, and in particular because our eyes have so much water in them, vision would be impossible if it weren't for this tiny crack in an otherwise long and impenetrable wall. Only by the dumbest of luck can we see, and only by the dumbest of luck can light get far enough into the ocean for photosynthesis and thus for life to occur. Just a tiny shift in the optical properties of water, and life as we know it would not be possible anywhere in the universe.

Moving back from the cosmic, water is not equally transparent to light even within this tiny visible range. Water turns out to be far more transparent to blue light than to light of any other color. Water is fairly transparent to green light, but strongly absorbs purple, yellow, orange, and especially red light. We don't often notice this, because the bodies of water we look at are too small (e.g., a water glass or bathtub) or too full of other things that absorb light even more strongly (e.g., the mud in the pond at my farm). Even most beachgoers don't see this, unless they go to the tropics where water is clear. You do see the effect in a pool, though, if you look straight ahead. The far wall will look blue. But even this can be hard to realize, because pool walls are often painted blue.

These differences in how different colors travel through water pile up when you have a lot of water. This is because the light level is an exponential function of depth rather than a linear one. Certain things are linear functions of depth. Water pressure is a good example. If the pressure at 30 feet depth is about 1 atmosphere (this excludes the 1 atmosphere from the actual atmosphere above the water), the pressure at 60 feet is 2 atmospheres, the pressure at 90 feet is 3 atmospheres, and so on. You can say that the pressure in atmospheres is just the depth divided by thirty. Light doesn't work like this. Suppose that the amount of blue light at the surface is 100 in some unit. Suppose again that the blue light level at 30 feet is 50. You might think that the light level at 60 feet would be 0, but instead it's 25. And at 90 feet the light level is 12.5. So instead of subtracting a certain value every time you drop 30 feet, you instead divide by two. We call this an exponential relationship, and it has a huge effect on the color of light as you go deeper. Suppose that in this same ocean, the amount of red light is also 100 at the surface, and at 30 feet is 25. This means that red light drops fourfold for every 30 feet of depth. So, at 30 feet blue is 50 and red is 25, making blue light twice as bright as red light. At 60 feet blue is 25 and red is only about 6, so now blue is now about four times as bright as red. At 90 feet blue is 12.5 and red is 1.5, making blue about eight times as bright as red. This process continues, and eventually the blue light is millions of times brighter than the red light, and also far brighter than the orange and yellow light. Green light holds on a bit longer, but once you get to about 100 feet or so in open ocean waters everything is blue, eventually becoming laser-like in its purity. So, in the end, the colors I never forgot from my first submersible dive are due to small differences in how water treats different colors combined with some math and a lot of volume.

This exponential relationship affects the visual world of pelagic animals, which affects how they find food and one another, and how they hide. Before we get to this, though, we need to acknowledge that water isn't the only thing in the ocean. In the clearest and emptiest parts of the ocean, like the Sargasso Sea at the center of the Atlantic, the water has almost nothing else in it. However, other parts of the ocean contain phytoplankton. These are single-celled photosynthetic organisms, some of which are related to plants, others more closely related to bacteria. Both types strongly absorb blue and red light for photosynthesis. They mostly leave green light alone, so their presence makes the ocean greener. As we move farther away from the tropics, these phytoplankton become more abundant, so the water becomes greener. The phytoplankton are also more common near most coasts, which makes the water greener still. In addition, coasts are where rivers run into the ocean and add large amounts of dissolved and undissolved organic matter (i.e., dirt) to the mix, which tends to make water browner. The organic matter and phytoplankton absorb far more light than pure salt water, so their amounts have an outsized effect on how murky the water will be and what color it is. Going back to chapter 1 and my trips to the beach, this is why the ocean as seen from the beach looks nothing like the bulk of the ocean offshore. The coastal waters are the unhappy families of *Anna Karenina*, each different in its own way and for its own reasons. Moving away from these to the "happy" open ocean, we now explore how pelagic animals see in this blue and often dark world.

## The Big Squeeze

A frequent memory from my childhood is practicing little league baseball in a field across the street from my house. We'd start practice after school at three thirty p.m., stopping at five thirty because in Pittsburgh in autumn it got pitch dark by about six. At the start of practice we had full sunlight (assuming it wasn't yet another dreary day), and by five thirty it was too far into twilight to play safely, but still bright enough to do most things without needing streetlights. I often ask students in my classes how much darker the end of the practice was than the beginning. Ten times? A hundred? A thousand? It turns out that the end of practice was one hundred thousand times darker than the beginning. There is really nothing else in our lives that changes as dramatically as the light level, but we seldom recognize the extent of the change. We all have noticed light levels dropping rapidly right before a thunderstorm, but few of us realize that they've actually dropped up to a hundredfold. We all know that that a moonless night is quite dark (though you can still get around in it), but it's hard to believe that it's ten billion times darker than a sunlit day. This is true for two reasons. First, our eyes adapt partially to darkness, which we'll talk about soon. More importantly, though, we compress this enormous range of intensities into a manageable package using what is called logarithmic sensing. For those who didn't love math in school, logarithmic is the opposite of exponential—where things change very slowly rather than very rapidly. More precisely, if a light value is 100 and we perceive it as a "one" in our minds, a light value of 1,000 is perceived as a "two" and a light value of 10,000 is perceived as a "three." This takes the ten-billion-fold range of brightnesses from a moonless night to full sunlight and converts it in our eyes and mind into a much smaller range of perception.

This convenient sensory compression, which I'm calling the big squeeze, is found not only in our eyes. It's also in our ears, both in pitch and loudness. For example, each octave in Western music doubles the pitch of the one before it, so the highest key on a piano has a pitch that is 150 times higher than the lowest key. Loudness is measured using a logarithmic unit called a decibel, where every increase in 10 decibels is a tenfold increase in loudness. So a 105-decibel gas mower is not just a little louder than the 95-decibel mower next to it, but ten times as loud. We perceive it as only about twice as loud, but it's far more damaging to our ears, so shop carefully for mowers if you like your hearing. Even the intensities of earthquakes are described using logarithmic scales.

Getting back to eyes, another reason we don't notice the enormous changes in brightness over a typical day or night (or in a trip down in a submersible) is because we adapt to it. We often think that the changing size of our pupil does this, and it does indeed grow and shrink with illumination, but this doesn't come close to doing the job. If you're young, your pupil can grow to about 8 mm in diameter in the dark and will shrink to 2 mm in bright light. This fourfold difference translates into a sixteenfold difference in the amount of light that enters our eye. This is nice, but, remember, even just a heavy thunderstorm can drop the light level by a hundredfold. Most of our adaptation to dimmer illumination is done by the changing of the sensitivity of the cells in our eyes, along with the switching eventually to a different set of cells, known as rods, from those known as cones. This switch from our relatively insensitive but color-discriminating cone system to our more sensitive but color-blind rod system happens at light levels around those of the full moon. This is why landscapes lit by the moon always look so silvery. If we could still see color at these light levels, we'd notice that the moon is actually brown and that the sky is royal blue. As we all know from stepping into a dark room at night, dark adaptation takes time, roughly thirty to forty-five minutes for us. Interestingly, this is also the length of twilight, so it seems that the dark adaptation rate is set up to keep pace with the rapidly dropping light levels at the end of the day.

The other thing that occurs as we switch from the cone to the rod system is that our visual acuity falls apart. For example, most people are unable to read under the light of the full moon. This can be frustrating if you are a night hiker, because although it is more than bright enough to get around, you can't read the map. This happens because we neurally tie about ninety neighboring rod cells together at a time for each "pixel" of our vision. Because each pixel is now ninety times larger, it's ninety times more sensitive, but this ruins our ability to see fine detail. All these things matter when we think about the visual world of pelagic animals.

## How We Study Vision at Sea

Pelagic animals have to deal with increasing darkness not only as they go deeper but also as day goes to night. Night is obviously dark everywhere, but night underwater can be intense, even at shallow depths. I've done a reasonable amount of night diving. We'd of course use flashlights, but now and then I'd turn mine off, and even at depths of only 60 feet the darkness was profound, as at the times I've turned off my light in a cave. It's so dark that the lack of light feels like a physical presence. Pelagic animals also have to deal with the intense blueness of the illumination and the lack of all other colors. Together, the darkness and the blueness, and how they're affected by depth, create an interesting challenge for the animals in the ocean. One might hypothesize that they've evolved various strategies to become as sensitive as possible, and that they are more sensitive to blue light than light of any other color. The trick is how to find out.

I've worked on many things over the years, but my main interest has been understanding the visual world of pelagic animals, and how they deal with it. I do little work on vision itself, but am nearly always together with vision scientists when I go on a research cruise. In particular, I often go to sea with Tammy Frank, who is I think is the greatest open-ocean vision scientist of her generation. There are multiple ways to study vision. Some researchers examine the chromosomes of animals to see what genes they have for vision compared to those found in other animals. Others pull individual rod and cone cells out of the retina and put them under custom-made microscopes that measure how well these cells absorb different colors of light. Some do behavioral work, having animals chase around targets of different colors and brightnesses. Tammy's specialty is electroretinography, also known as ERG.

ERG relies on the fact that our nervous system is a bioelectrical computer and thus generates electrical signals that can be measured. ERG is helped in particular by the fact that the retina is a dense collection of neurons that all fire off an electrical signal at the same time if they're exposed to light. The collective signal is so strong that it can be measured by an electrode that just touches the eye. So, in theory, if you want to find out how sensitive an eye is to light, you just shine a light of ever-increasing brightness into it and measure the electrical response. And if you want to know the sensitivity of the eye to different colors of light, you shine different colors into it and again measure the response. But, to paraphrase the British comedy show *Monty Python*: playing the flute is easy; you just blow into this hole and run your fingers up and down. ERG turns out to be a fussy technique, even on land. At sea, it becomes almost comically difficult because ships roll, rock, lurch, and endlessly vibrate, so doing ERG on a research cruise is much like threading a needle over and over on a multiday roller-coaster ride. To make matters worse, everything has to be done in the dark. Because we typically have only one research lab on the ship, which has to be shared by many people, and because this lab typically has only one light switch, creating darkness involves making small, low-budget darkrooms out of rolls of trash-bag plastic. We first set up the various research gadgets on the counters in the lab after schlepping it all on board—everything ocean-related seems to be heavy. We then drill holes into the wooden counters (assuming the ship's captain lets us), and drive in screw eyes that create anchor points. We use these and a crate of bungee cords of every length to strap down all this expensive equipment so that the first big roll of the ship doesn't toss it all on the floor. Then we get out large rolls of trash-bag plastic, cut them into carpet-sized sheets and do our best to wall off the portions of the lab that need to be dark. Important considerations are avoiding overhead light fixtures and portholes in our "dark room," but making sure to include an air-conditioner vent so that the hot equipment doesn't create a mini sauna. The plastic is attached to the ceiling and floor using rolls of duct tape, and an overlapping shower-curtain-style "door" is made at the end. Once made, we get inside and dark-adapt, only to discover that we have dozens of light leaks, which leads to another hour of patching these with more duct tape. Once completed, one of these dark rooms becomes Tammy's home for the entire cruise. She does come out to eat, but I'm not sure she sleeps. So, if you ever read a news article about what some animal can see in the ocean, you can bet that it's been done by a person living off candy bars for two weeks inside a trash bag with a counter of homemade equipment.

This all, of course, assumes that everything works. Research at sea can fail for any number of reasons: the boat stalls, the winches jam, the trawl net won't open, the trawl net won't close, there are no animals, or the prima donna ERG equipment goes haywire in a host of different ways. As I said before, at sea plan A is a joke, plan B requires a miracle, and plan C will fail in a way you'd never expect, like the prong snapping off the electrical plug of a critical piece of equipment. So, a couple days into the cruise you're left with plan D, which typically involves even more duct tape, hose clamps, cannibalized ship parts, and a lot of creative thinking and good humor. On one cruise, my postdoc Sarah Zylinski was left shining colored flashlights at deep-sea octopi in the bowels of the ship near the engine because the rest of her gear broke. Amazingly, that worked and led us to a new way of thinking about camouflage in the ocean, so plan D isn't always bad.

Not everything we do in open-ocean research looks like a kitchen remodel with too few trips to the hardware store. We also do some high-tech stuff, like genomics, microscopy, and mathematical modeling, but we seldom do this while we're actually at sea. Mostly, we collect animals, preserve them in various ways, and bring them back to shore. The primary reasons for this are the logistical hurdles I already mentioned, but another is that we all get a little dumber offshore. We don't know the reason for this—my former advisor Edie thinks it's because we're using so much of our brains just to stand upright on a rolling deck—but it's hard to be clever at sea and easy to make mistakes.

In sum, getting information about pelagic animals is difficult, but worth it because so little is known. So let's turn now to what we do know about how animals deal with the darkness and blueness of this habitat.

## How to See Underwater; or, "Everything Is a Trade-off"

Before we get to the adaptations that are specific to the pelagic ocean, we need to know how these animals see at all. Nearly all of us have opened our eyes underwater, and the first thing we notice is that our vision is blurry. The reason for this is that the sharpness of our vision is not primarily due to the lens in our eye. Over three-quarters of the focusing power of our eyes is due instead to our cornea, the window at the front of our eyes. Because this window is curved, and because—like glass—it has different optical properties than air, it focuses light. Our lens is mostly used to adjust our focus for different distances so that we can, for example, see both the road and the speedometer when we drive. It does this by changing shape, a process we don't appreciate until we get older and our lens—like our joints—becomes stiff. At this point, we're no longer able to see the speedometer or even read a book without glasses.

As we discussed in chapters 2 and 3, our tissues are made of mostly water. For this reason, their optical properties are fairly similar to those of water. So when we dive underwater, the cornea is no longer able to focus light. This leaves us with only our lens, which can't do the job. It's an impressive failure. If your vision is 20/20 on land it will be about 20/650 underwater. This means that something you could see from 650 feet away on land you'd be able to see only from 20 feet away underwater, which is why we wear face masks. This isn't just a human problem; many terrestrial animals use the cornea as their primary focuser, and will have dramatically worse vision underwater.

Many pelagic animals have eyes just like ours though—fish, squid, octopuses, and even some worms and snails. Do they all live in an unfocused world? It turns out that they solve this problem by not using the cornea to focus light. Some of these animals don't have a cornea at all, and those that do have flat corneas that couldn't even focus light on land. Instead, they invest everything in making a more powerful lens. The strength of a lens depends on how curved it is, going from a flat lens that does nothing at all to a spherical lens that focuses light as strongly as possible. Our lenses, and those of many terrestrial animals, are like weak magnifying glasses. The lenses of fish, squid, octopuses, and these snails and worms, however, are shaped like perfect balls. They're also dense. Our human lenses are rubbery (at least when we're young), but the center of a fish lens is like a little rock. This extra density affects their optical properties and gives them even more focusing power, enough to compensate for what a cornea can do in air.

It's a wonderful evolutionary adaptation for underwater vision, but as with nearly every adaptation in biology, it can lead to new problems. The first problem is that the lens can no longer change shape to adjust the focus. These animals need to be able to see at multiple distances just like we do, so instead of changing the shape of the lens, they move it backward and forward in the eye, which is how we focus camera lenses. A second problem is that a ball makes a terrible lens. If you've ever looked through a clear marble, you'll notice that everything is blurry and distorted. These animals solve this problem by having lenses that are layered like those spherical candies that change color as you suck on them. The very center of the lens is the most dense and focuses light the most strongly, but as you move out from the center, the lens material become less dense. It does so following a specific mathematical relationship, which in the end creates a ball lens that focuses light perfectly. This is yet another example of the fact that nearly every clever thing invented by humans was evolved by animals countless years before. It's only fairly recently that optical engineers have been able to make lenses of this sort, and I would guess that they were inspired by fish lenses.

Some animals want to have sharp vision both under and above water and can't just put on a face mask like we do. The four-eyed fish (*Anableps*) does this by having permanent bifocals. Each eye is divided into upper and lower halves, with a football-shaped lens at the center. When the fish sits at the water surface (which it often does), the light from above water goes through a curved cornea that focuses most of the light, and then through the shorter, less-curved axis of the football lens. The light from below the water's surface goes through a flatter cornea and the longer, more-curved axis of the football lens. Using this clever system, the fish can have both a terrestrial and an aquatic eye using just one lens. Some diving birds that need to be able to see prey underwater use a different system. Their lens is especially rubbery, owing to the presence of various molecules inside it. When the bird dives, the muscles that normally change the shape of the lens actually push it through the pupil, squeezing the front of the lens so that it has a much more curved front surface and thus acts more like an underwater ball lens.

These adaptations allow pelagic animals to focus light, but how do they deal with the darkness at depth and at night? As

I mentioned, it can be reasonably bright at depths of 600 feet in the warm tropical ocean, about the same as a heavily overcast sky, but much of the ocean has colder water. Colder water typically is more "productive," which means it has more phytoplankton. As mentioned earlier, these single-celled algae can grow in tremendous numbers and make the underwater world much darker. And, of course, everything is darker at night, especially underwater. So even non-deep-sea pelagic animals have to deal with low light levels. As we've already discussed, changing the size of your pupil, even assuming you can, isn't close to enough. So, although a number of pelagic animals do have large pupils, they need to do more, and seem to primarily use four additional tricks.

First, many of the animals—especially the deeper ones—give up on color vision and have a monochromatic visual world that is most sensitive to the brightest color in the water, which in the open ocean is blue. Second, some of these animals have very thick retinas that are also backed with a mirror called a tapetum. This means that not only does light have a longer trip to the back of the eye, but it gets a second trip as it bounces off the mirror and goes back up through the retina again, increasing the chances for it to be detected by the rods and cones. All this light bouncing around causes extra glare and thus reduces the quality of the image, but it is apparently a worthwhile trade. Tapeta are seen in land animals too, especially ones that come out at night or during twilight. Autumn in North Carolina means deer are all along the roads back to my farm. No one wants to hit a deer, but they can be hard to see, so the Department of Transportation tells us to look for their eyes reflecting our headlights.

The other two tricks are called spatial and temporal summation, and this is where more trade-offs come in. We already mentioned that human rods are tied together in sets of ninety, which increases their sensitivity but costs us the ability to see detail. This is the spatial part of summation and it is practiced to an even greater extent in pelagic animals, especially in those at greater depths. Many photoreceptors can be tied together to make quite large pixels that will collect more light, just as a trash can will collect more rain than a paint bucket. It's a good solution for low light levels, but the bigger your pixels are, the fewer of them you can put in your eye, which will make your vision coarser.

You might think, why not just use small pixels and just "turn up the volume" on each pixel? That way you could see your meal in the dark and recognize it too. Neurons are actually good at amplifying a small signal. Unfortunately, you can't do this with light, because light has the annoying characteristic of being packaged in chunks, called photons. Although we might think of light as some smooth ethereal fluid entering our eye, it's actually more like a sloppy rainfall, with chunks of light falling randomly across the backs of our eyes. The rush of a heavy rain sounds smooth and steady to us, but in a light rain you can hear the individual drops, which seem to fall randomly. Seeing in the dark is much like counting raindrops in a light rain. The areas under the trees will have fewer raindrops than the open areas, but you won't see a clear "shadow" of the trees reflected in the pattern of drops on the ground. While daylight is analogous to a thunderstorm of rain, we actually see this individual raindrop effect when we walk outside on moonless nights. That shimmering you see is partially due to the randomness of individual photons striking the cells of your retina. If we were to just amplify this signal in our brain, the scene might look brighter, but we wouldn't get more information because it would still look like a collection of random events.

So, pelagic animals are stuck with averaging what they see over space, and likely use just enough of it to balance the loss of information due to what we call "photon noise" with the loss of acuity from having overly large pixels. This leaves many of them with crude vision—in fact, worse than what we call "legal blindness," which is 20/200 vision. One way for these animals to increase both the sensitivity and the sharpness of their vision at the same time, though, is to increase the size of their eyes. If you double the area of your retina, you can double the area of each pixel to increase sensitivity and still have as many pixels. This comes with its own problems; the first one being that you can only cram so much eye into a head. In fact, if you look at cross-sections of the heads of many pelagic animals, you'll see that their two eyes nearly touch in the middle. Eyes are also energy hogs, being extensions of the brain, which is the biggest energy consumer for its size in our body. Some pelagic animals, especially the deeper ones, solve these problems by making the eyes tubular instead of spherical. So they get the advantages of a big pupil and lens, but at the expense of not being able to see in all directions.

Temporal summation is a different solution. Instead of making the pixels larger, you expose them to light for a longer time. This is a lot like changing the shutter speed in a camera, but even more like changing the frame rate in a movie camera. A slower frame rate gives each pixel a greater chance to catch what light is there, making it better able to see in the dark. It should come as no surprise by now this comes with a cost. The cost in this case is that a slower frame rate makes it harder to track fast-moving objects. If you ever watch an indoor sporting event, you'll notice that the press photographers are using cameras with enormous lenses. The lenses are often several times the weight and size of the camera itself. The lighting in indoor sporting events is often bright, though, so why do they need such huge lenses? It's because athletes tend to move quickly and don't like flashes going off in their faces while they compete. So the cameras need to have short shutter speeds to freeze the action without a flash, and a huge lens to capture enough light during that short time. So pelagic animals working in low light are sacrificing their ability to see both detail and motion in order to see at all. In sum, seeing in the dark is a hard problem, and like all hard problems, each solution leads to more hard problems. Still, there is another solution: bringing your own light.

## Bioluminescence

As I mentioned, my first home of Pittsburgh was not a nature preserve, and as a Cub Scout I visited more factories than forests. But we did have fireflies. In late summer, our twilight yards were filled with them, and seeing them as a young child gave me my first sense of magic. My current home has far more fireflies—the summer pastures and trees are filled with hundreds to thousands of them, each doing the slow blink and slide that tells the others, "I'm here." Many nights, one will attach itself to the screen of our bedroom window, flashing lazily. We've never quite understood why, but think it's trying to communicate with the blinking light on my wife's alarm clock.

Given my fascination with this phenomenon, I was excited to start my postdoc with Edie, who was an expert on marine animals that made their own light. As I said in chapter 1, she would answer each call with the word "bioluminescence." Being young and obnoxious, I'd usually answer her with a random word like "pork chop," but secretly I shared her passion. My first experience with marine bioluminescence was going to the bow of the ship on the night of my first cruise with Erin Fisher, an undergraduate in Edie's lab. She and I leaned as far over the rail as we dared and watched as various plankton and algae flashed as they were churned up by the turbulence of the wake. The light was eerie, but also so dim that I couldn't even see it as a color. It wasn't until we had done our first trawl that I saw the light up close. We had pulled the 100-foot-long net on deck and detached the six-foot-long opaque container at the end of it that held all the animals. This container, known as a "cod end," was full of water as well, and it took two of us to lug it up the slippery, rolling deck of the ship to the lab. Once in the lab, we locked the door, turned off all the lights, dark-adapted for a few minutes, and opened the cod end, allowing the water and animals to pour into a large plastic tub on the floor.

I've been trying to reach for a metaphor here, but sometimes something is so new that it exists only as itself and can't be compared to anything else. All I can say is that dozens of animals poured out of the cod end, seen in this pitch-dark room only because they were all glowing by their own light. Some flashed so rapidly and dimly you weren't quite sure you saw them at all. A few flashed very brightly. Some just turned on and wouldn't stop. A few blew apart into a million shining pieces, and one or two were sending pulsating waves of light up and down their bodies. What united them all was their color. Each was a stunning and rich blue, like a tiny, brilliant piece of the blue world I saw when I later looked out of the window of the submersible.

Several years later I was doing deep-sea light measurements in the Bahamas with Edie. This involved making the submersible neutrally buoyant at various depths and then taking thirty-minute-long measurements with a large gadget that was in the back chamber with Edie. It was tedious, made worse by the fact that I had thought the dive had been canceled, drunk two cups of coffee before we went down, and desperately needed a bathroom. I did my best to distract myself by looking outside the sphere for signs of light and life. At the shallower depths, I did see some animals swim or float by, but at no depth did I see even a spark of bioluminescence. So how does one reconcile this with the glowing fountains that came out of the cod end?

It turns out that the majority of bioluminescent organisms hoard their light and will glow only when touched. There are good reasons for this. First, as with any recipe, it takes energy and ingredients to make light. In the case of bioluminescence, the light is made by adding oxygen to a small molecule known as a luciferin. There are about seven to nine known kinds of these molecules among the various organisms that glow. The reaction is sped up by an enzyme called a luciferase, of which there are many types (the terms luciferin and luciferase are stolen from firefly researchers, who themselves stole them from Lucifer—the light bringer). In a few cases, this reaction requires the addition of another molecule called ATP, which adds energy to get the reaction going. The luciferins themselves are also costly to obtain or make. The second reason, which I feel is more important, is that the light will give you away. The pelagic world, especially at night or at depth, is utterly featureless. It's also fairly empty of food, at least compared with a coral reef or forest. Light means life, which means food, so the moment you flash you have attracted the hungry attention of everyone around you. So you had better be flashing for a good reason. The situation always reminds me of those submarine movies, where two submarines are trying to be as quiet as possible so that they don't give themselves away to each other's sonar—and then someone drops a wrench. Except in this case, the area is filled with countless tiny submarines, each waiting for someone to flash and give themselves away.

Given this, you might think that bioluminescence, while stunning, is a terrible idea. It appears to have evolved at least forty times, though, primarily in the ocean, so there must be a good reason for it. And this is where things get tricky. Oceanic research can be wonderful and exciting, but one of its great limitations is that it is exceedingly hard to get observations of undisturbed behavior. For example, if we want to know all about the social lives of beavers, we can set up tiny Wi-Fi-enabled cameras in their territory and simply watch them. In the pelagic ocean, this is much harder because the habitat is immense, accessing it is difficult, and animals are often sparsely distributed. On top of this, just about everything you put into this habitat sticks out like a sore thumb, bringing in new animals and changing the behavior of the animals that are already there. As I mentioned earlier, part of a blue-water dive rig is a float that is about 2 feet across—ours is bright red. We typically dive for about thirty to forty-five minutes, and nearly every time we surface we see little fish hanging out under the float, even though we didn't see a single fish on the dive. Pelagic animals, who spend their lives in the liquid version of outer space, truly hate being in captivity, so we can't just collect some and study their behavior on the ship. So, studying the behavior of pelagic animals is like quantum mechanics: the second you try to observe you disturb everything. We are exploring ways to get around this, with some success, but it's not easy.

This is a roundabout way of saying that, although we have many ideas about what functions bioluminescence serves in the pelagic world—some of which we're fairly sure of—we have direct behavioral evidence for very few. Instead, we do what I like to call forensic science. We bring up the animals, study their eyes and light organs in the context of what we know about the depth they live at, whom they might interact with, and other factors, and come up with our best guess about what is going on. For example, imagine you see a beaver (I like large rodents) under a fallen tree that looks like something has chewed through it. Then you notice that the beaver has teeth of the right size and shape to make the chew marks on the tree. So you conclude that the beaver chewed down the tree, even though you never saw the beaver do it.

So what are the things that we think bioluminescence is doing in the pelagic ocean? Most seem to involve either finding and catching food or avoiding becoming food. Since we'll cover this all in more detail in chapter 6, I'm only going to briefly mention them here, starting with the animals that glow without being touched. First, we're pretty sure that some are using their lights as lures. Anglerfish in particular are wonderful examples of this, with a glowing lure on a rod that hangs right above their mouth. Others, especially certain fish and shrimp, have light organs directly under or in front of their eyes, which seem to work as flashlights that they use to find prey. A third use involves covering as much of their belly as possible with lights that match the brightness of the light coming down from above. This, known as counterillumination, hides their silhouette, which is what we think it is used for. Finally, there are a number of free-floating bacteria that, if they sense the presence of other bacteria, glow dimly but continuously for reasons we don't understand. These bacteria have been co-opted into the light organs of some fish and squid in a symbiotic relationship. The host provides protection and certain resources, and the bacteria provide light.

The rest of the bioluminescent events seem to occur only when the animal is touched. Our best guess is that these flashes of light serve various defensive purposes. Imagine you are a small bioluminescent organism in the dark and a potential predator comes up and starts poking you. In what ways could you use your light to help you? First, if your light is bright enough, you might temporarily blind the predator. We've discussed dark adaptation, and we all know that painful feeling of stepping out into bright light after watching a movie in the daytime. The predator may take a while to recover from this, giving you time to move away. Or you could use your light to distract the predator. A pelagic worm known as the "green bomber" does this—releasing up to eight bioluminescent sacs from its body that can distract a predator while the worm itself escapes. A third trick is to have sticky bioluminescent slime that coats an animal if it touches you. This is a bit like the purple dye some banks have in money bags during a robbery, which later explode covering both the money and the robbers in ink. In this case, though, the ink is glowing, making the predator now a target itself. Finally, your own flashing might attract another animal that is even larger and more dangerous than the predator that is bothering you, with the hope that this uber-predator will attack your predator, leaving you again to run off. This tactic, which reminds me of too many Godzilla movies, is called the "burglar alarm," and is contested among bioluminescence researchers. There is some laboratory evidence that this might occur, but no one has ever seen anything like this in the wild, though we do know that predators will attack bioluminescing animals. It's also always seemed dicey to me, since for all you know you're just going to attract more animals that are interested in you, or the uber-predator will eat you after it eats the lesser predator.

It will take some time before we know which of these functions turn out to be true for which animals, but overall it does look like bioluminescence primarily functions in ways that help them either catch prey or hide or escape from predators. But you might say, "What about communication?" After all, the bioluminescent system we know most about, that of fireflies, is all about communication. This likely does occur in the pelagic ocean, but it may take time for us to discover it. If pelagic behavior is hard to study, pelagic social behavior is even harder. We do have one remarkable example in shallower water, though—the bioluminescent ostracods.

Ostracods are small shrimp-like creatures that build a little hinged double shell around themselves, likely for protection. Some are found floating in the pelagic world, but most are found closer to the bottom and in relatively shallow water. They're quite diverse, with at least thirteen thousand species. Most are small, but a few can be as large as a marble. Some make their own light, often presumably using it for protection in the ways described above, but a few use their light for mating. We'll talk more about this in chapter 7, but finding a mate in the ocean can be hard, and finding a mate in the dark ocean can be even harder. So, certain ostracods let others know what species they are by marking out a light dance in time and space. They will move through the water in a certain path, and flash at certain times along it. Five flashes shaped like an "L" mean one species; eight flashes in a spiral mean another. They also, like fireflies, will sometimes respond to a flash with a flash of their own, which leads to one of the most remarkable things I've seen.

About eleven years ago, I was working with various colleagues at a field station on a small island on the Belize Barrier Reef. This island was only half an acre in size, with about twenty palm trees, a few simple buildings, and a hole cut in the dock for a toilet. The island was just a sand pile on top of a reef, and in a way as much a research vessel as the ones we normally use.

Instead of it moving through the water, the currents moved water past it, always giving us something new to look at. For the first few weeks we were in the dry season, which meant a howling wind and no rain. One night in the fourth week we passed abruptly into the wet season, and from then on had eight to twelve hours of biblical storms, starting around dinnertime. I had lived at the top of a high hill in Pittsburgh, and then in the southeastern United States, so I thought I was used to thunder and lightning. But I had never seen anything like this. One stormy night, we were all sitting on the balcony of the main research building, looking at a shallow area of the reef that we knew had bioluminescent ostracods. After a few minutes, we noticed that every time the lightning flashed, about a hundred ostracods would flash back a second or two later. We watched this exchange for over an hour, none of us saying a word. As a kid I loved Greek mythology, and would read a book I had about it over and over. The first page had a drawing of Uranus the sky god, his face filled with stars, looking down at the earth goddess Gaia, who was looking back up at him with stars in her eyes. Forty years later, I sat on a tiny island with my friends, watching the sky and the sea talk to each other.

# CHAPTER 5

# Motion

> She kept swimming out into life because she hadn't yet found a rock to stand on.
>
> —Barbara Kingsolver, *Animal Dreams*

## Going from Point A to Point A

I have had a number of odd and oddly specific goals in life. One that has been with me since I was a child playing in the North Carolina surf was the desire to be out of sight of land. I'm not sure where this urge came from. Maybe it was my native introversion, maybe a need to escape the claustrophobic streets and hills of Pittsburgh, but my best guess is that it was an extension of my childhood urge to swim as far from shore as possible—an urge that nearly killed me once or twice. I finally got my wish on my first cruise, where I spent much of my time on the highest open deck of the RV *Edwin Link* looking at all the nothing in every direction. It did not disappoint.

The scale of the open sea is of course immense. The 02 deck (as they call the deck that is two stories above the main deck) is 25 feet above the water. At this height, the horizon is about 6 miles away. So, you are looking out at an area that is over 100 square miles in size, which is about double the area of Pittsburgh (or five times the area of Manhattan, for those who don't visit the Rust Belt). Except for the clouds and waves, there is nothing to see. At the same time, this view seemed small and personal. This latter sensation is harder to describe, but the complete lack of features made me feel like I was in a pocket universe, bounded by a hemispherical dome of sky that was fastened to a circular ring at the horizon.

Completing the illusion was the absence of any sensation of forward motion. As I've mentioned before, our ships go about 11 miles per hour. This is hardly drag racing, though it would be a fast run. But in the ocean, with no reference points, there's no sense of forward motion at all. The engines grind away, and the propellers leave a churned wake that may go on for miles, but there's no sense of going anywhere. During transits, we often sit at the bow of the ship on top of enormous lockers, which hold equally enormous ropes that can tie the ship to a dock. There, the engine noise is absent, and we can't see the wake, so there are truly no motion cues. Transiting the open ocean is much like riding an elevator. You step inside, nothing changes, and then you step out in a new place. The only difference is that the elevator is 100 square miles in area, and you can be inside it for weeks.

Our 100-square-mile elevator is, of course, tiny compared with the size of the ocean. In fact, it's less than one-millionth the area. The immensity of the oceans and our relatively slow passage through them have practical implications, the biggest one being that we try not to go farther from shore than we have to. First, the amount of fuel required to drag a 200-foot-long vessel to the center of the Atlantic Ocean (let alone the center of the Pacific) is beyond ridiculous. Large cargo vessels measure fuel consumption in tons per hour. Our ships are much smaller, but the fuel consumption is still large. Second, because our ships are slow, medical help can be days away if something goes wrong.

The ocean can be dangerously large even when we're close to land. In 2008, I did some scuba-based research on Heron Island. This is a coral island that is part of the Capricornia Cays—a cluster of small islands that are the southern boundary of the Great Barrier Reef of Australia. Many of the cays are exposed only at low tide, making these lonely half-submerged islands both beautiful and eerie. One day, we were diving right off Wistari Reef, which was a couple of miles from Heron Island. We surfaced after the dive, and our boat wasn't there. It wasn't anywhere on the horizon, either—at least from our viewpoint just a few inches above the water. The leader of our group, Justin Marshall, was an experienced diver who had worked at Heron Island for decades. He was (and still is) an unusually calm person. He reassured us that the dive boat "would likely show up soon," but after fifteen minutes of no dive boat, we noticed that we were drifting away from Wistari Reef. We were drifting toward Heron Island, which was good, but it was clear that we'd miss it by at least a mile, which was not good. I asked Justin what came next after Heron Island in the direction we were drifting. "Antarctica." So, it was decided that we should inflate our safety sausages, which are industrial-strength 6-foot-long red party balloons that are easier to see from afar than a set of wet, bobbing heads. After another nervous half hour, the mast of our dive boat appeared over the horizon, and eventually came to pick us up.

All this is to say that the open ocean is impossibly large and featureless, to the point where you have to ask, "What is the point of going anywhere?" Everywhere looks like everywhere else, so even if you know where to go and have the ability to cover the distance, would the trip even be worth it? What ultimately drew me to biology was its existential humor, and one of the things that drew me to pelagic oceanography in particular was the ridiculous mismatch between the size of the habitat and the animals within it. I once saw a tiny juvenile fish while diving over a deep-sea canyon far east of Cape Cod. The fish has no common name, just the Latin genus *Pristigenys*, and the one I caught was only half an inch long. But it was superbly orange, with a pleated silver dorsal fin with orange spots, each spot lined with a perfect black rim. The spots were so small that you almost needed a magnifying glass to see them, but they were as perky as the corsage on the tuxedo of a teen going to the prom. That night after dinner, I kept thinking about that fish. Who were the spots for, and would anyone ever see them? What was such a tiny and perfect thing doing in such an empty world?

This chapter is about motion in the pelagic world. It talks about how the animals move, but also why they move and how they know where they're going. It turns out that many animals do have reasons for going from point A to what looks like point A, and the ability to see signs invisible to us that take them through this apparently empty world.

## The Swimmers

Many open-ocean animals do swim as we typically imagine it. Fish swim, as do sharks, rays, sea turtles, and marine mammals. Fish swimming has been divided into at least ten types, with exotic names like "anguilliform" and "thunniform." When I was in graduate school there was a fish lab down the hall full of students who would argue endlessly and loudly about which fish was using which type of locomotion. Fish locomotion is such a large and sophisticated field that I hesitate to say anything about it, except that the various locomotory modes are primarily about which parts of their body are stiff—and thus contribute little to thrust—and which are flexible, and thus provide most of the thrust. For example, in anguilliform locomotion, which means swimming like an eel, the entire body undulates roughly equally from stem to stern. In other words, these fish swim much like a snake moves over grass. In thunniform locomotion, which means swimming like a tuna, the body is so stiff that nearly all the thrust is provided by the sideways undulation of a large and typically crescent-shaped tail. Some fish do not undulate their body or tail at all, and move entirely by oscillating their fins, though this is more common among reef fish, which have shorter distances to travel.

As with flight, different modes of swimming provide different levels of maneuverability and stability. Some fish pivot and dive like fighter jets, some lumber along like 747s. Most are somewhere in between. Overall, one can think of fish locomotion as analogous to bird flapping flight, but because fish are typically neutrally buoyant, they do not have to use their motion to keep from sinking. Some sharks are negatively buoyant, though, and generate lift from their side fins to keep from sinking, much like birds and airplanes do. And some open-ocean fish truly fly—sort of.

Flying fish, which are common in the Gulf Stream off the east coast of the United States, do look as though they are flying, with their pectoral (side) fins stretched out like wings as they scoot and skip dozens of yards from wave cap to wave cap. But these "wings" don't flap, and if you held a flying fish and let go, it would fall to the ground. What provides the forward thrust is an intense vibration of the lower tip of the tail, which remains in the water. Remarkably, this vibration is enough to get the fish up to a speed where it can take off and glide. I'm not sure how they pull it off; it'd be like one of us trying to run across a pool—and take off—by vibrating our toes in the water. Because they can't continue to power themselves after takeoff, though, they can't go far—maybe 100–200 feet in flights that last 20–30 seconds. They also can't get very high up in the air, with their flight height depending mostly on where in the wave cycle they leave the water. If they take off from the peak of a tall wave, they can get about 20 feet above the trough (bottom) of that wave, which can put them on the deck of a passing ship. They often land on our research ships, where we can see that they come in both monoplane and biplane versions—like old crop dusters, except with eyes. They're abundant in many of the waters we work in, and many times I've seen group after group pop out of the front edge of a wave and glide just above the water until they smack into another wave peak.

As with many behaviors in the open sea, we don't know for sure why they occur, but the assumption is that the fish are gliding to dodge predatory fish. I've often wondered how successful this is, though. Their fish predators include some of the fastest swimmers known—tuna, sailfish, marlin, and swordfish. Measurements of the top speeds of these predators vary widely, and the upper numbers should be treated with suspicion, but speeds of 40–60 miles per hour are commonly reported. For comparison, the world record for the 100-meter freestyle in humans is about 47 seconds, which is less than 5 miles per hour. The flying fish fly at about 40 miles per hour, so if it was simply a way to get away faster, I think it would fail. More likely, the fish, by exiting and reentering the water, are doing their best to lose themselves among the shiny undersurfaces of the waves. Sadly for the flying fish, though, the flight itself makes them vulnerable to predatory aviators such as the frigate bird. The open ocean is the only place where you can be eaten by birds and fish at the same time.

Rays, skates, marine mammals, and marine reptiles locomote in a way that is similar enough to that of fish and sharks to be called swimming, but there are a few differences. The biggest one is that—whether they are flapping their fins, their bodies, their tails, or some combination thereof—these animals are flapping them up and down. Nearly all fish and sharks undulate their bodies and tails side to side, although if their pectoral fins are important to locomotion, those typically go up and down. There is no hydrodynamic advantage to either direction—the water doesn't care which way you wiggle in it. The answer to this dichotomy, which I first heard of as a child reading Rudyard Kipling's *The White Seal*, comes in several parts. Marine mammals undulate up and down because they evolved from land mammals that had vertebrae that bent best in that direction (try touching your toes by bending your torso to the left or right). Reptiles, however, do have vertebrae that bend side to side easily, and sea snakes and the marine iguanas of the Galapagos Islands do undulate side to side like fish. Sea turtles are a different story. They have a long and complex evolutionary history, but they ultimately evolved from animals that also had limited ability to bend their body, leaving them to locomote with their fins, which are evolutionarily related to our arms and legs. Rays and skates are thought to have evolved from bottom-dwelling sharks, and—via the tweaking of fewer genes than you might expect—now have bodies that are nearly entirely fin. The fins that became the body are the pectoral fins, which are on the side and thus go up and down. In theory, they could have developed a body that was all dorsal and ventral fins, and those could have oscillated side to side, but that didn't happen. It did happen in the ocean sunfish (*Mola mola*). This clunky fish looks like a turtle swimming sideways. It has a stiff, roughly circular body, with dorsal and ventral fins that flap side to side. In the end, how you swim in the ocean depends on both where you came from and chance.

## The Jetters

Many of the remaining ocean animals that can make headway against the currents (squid, octopuses, and a host of jellyfish and jellyfish-like animals) swim via jet propulsion. To my knowledge, jet propulsion is not found in any land animal. It's always dangerous to say "never" in biology, and maybe there's some odd insect that uses jetting, but the problem with jet propulsion in air is that to gain any thrust at all, you have to move the air very fast. Jet propulsion is the most basic form of locomotion because it requires only throwing stuff backward. You can throw anything—rocks, spaghetti, tennis shoes—and how fast you go in the other direction will depend only on the weight of what you threw and how fast you threw it. You're probably thinking that the last time you threw a tennis shoe you didn't start sliding in the other direction, but that's because the shoes on your feet are sticky enough to hold you in place. If you were on some wet and slippery ice, you could move around by tossing things.

The recoil of a rifle and the way a hose spraying water pushes back into your hand are both examples of jet propulsion. The first moves a light object very quickly, the other moves something heavier but more slowly. The problem a land animal would have is that air is much less dense than either a bullet or water. In fact, air is over eight hundred times less dense than water, which means that to get the same amount of thrust in air as in water you need to either move a lot more of it or move it MUCH faster. In this situation, eight hundred is a big number. For example, to match the thrust you'd get from water leaving your garden hose at 1 mile per hour, you'd need to move the same volume of air at 800 miles per hour, which is above the speed of sound. Humans use jet propulsion in air, but only because we can build machines that accelerate large quantities of air to very high speeds. The exhaust wake of a commercial jet leaves the engine at 375 miles per hour, and that of the primary engines of a Saturn V rocket leaves the engine cone at 5,400 miles per hour. Both engines are also enormous, so they're moving far more air than any animal could hope to. Now, of course, the average bird isn't trying to go 600 miles per hour or reach the moon, but even scaled down to bird-appropriate speeds, jet propulsion in the air doesn't work.

Underwater, it's a different story. Water's density means that you can move it more slowly to get the same thrust. The second advantage that aquatic jet propulsors have is that most of these animals are not much denser than water themselves. Unlike land animals, which often require heavy skeletons to keep themselves from being deformed by gravity (imagine your facial features without bones), pelagic animals do not need them for this purpose. Therefore, many pelagic animals are mostly water. Finally, unlike aerial or land animals, most pelagic animals are neutrally buoyant and thus don't need to use jet propulsion to hold themselves up as well as send themselves forward. On the flipside, water is much denser and more viscous ("sticky") than air, which makes moving through it using any form of locomotion harder. On balance, though, jet propulsion—for animals that can't make actual jet engines—is far more efficient underwater than in the air or on land, so it's not surprising that it's relatively common in the pelagic world.

I wouldn't agree with my physicist father's assessment of biology as "stamp collecting," but many biologists do like to categorize things. It's about the only way to approach the seemingly endless diversity in both form and function we find in nature—and it makes us feel good, like tidying up a library. Among pelagic animals, one can divide the jetters into four groups: cephalopods, medusae, salps, and siphonophores. The cephalopods, which are the squid, octopods, cuttlefish, and the nautilus, can all travel using jet propulsion, but in the open ocean it's the squid that do it best.

Squid have a famously odd body plan, but the two parts relevant to their ability to jet are their mantle and funnel. The mantle is a large conical tube that encloses most of the body. The mantle wall is almost solid muscle, and although the inside of the mantle cavity does contain the animal's organs, it leaves lots of room for water. During jetting, water leaves the mantle cavity via a small, tubular opening called a funnel. The squid can control the direction of the funnel and thus the direction the animal moves. If this funnel were also the only way that water could enter the mantle cavity, the poor animal would go in reverse every time it sucked in more water for the next burst. Instead, squid fill their mantle cavities via openings on either side of their heads. These openings are larger than the funnel, so the water enters more slowly and thus doesn't jerk the animals backward. Squid can eject water through their funnel with a lot of force, which lets them travel at speeds up to 20 miles per hour. The Japanese flying squid (*Todarodes pacificus*) can jet so hard that it can leap out of the water and rocket for distances of about 100 feet, ejecting water from its mantle cavity as it goes. This is impressive, given that the squid can be 2 feet long and weigh up to a pound. Using water for jet propulsion in air works very well, because water is so heavy, and air is easier to move through than water. When I was a kid, they sold a toy that was a small, hollow rocket that you half-filled with water, and then used a mini bicycle pump to pressurize the air in the other half. When you released the rocket from the pump, the air pressure would push the water out through a nozzle in the bottom, and the rocket could travel at least 100 feet straight up—and then get lost in a neighbor's yard or, worse, go through a neighbor's window.

As with cars, speed isn't the only thing that matters. You also have to consider how much it costs you to travel. In cars, we use "miles per gallon"; in animal locomotion, we use "cost of transport," which is how much energy it takes to travel a certain distance. In a way, you can think of cost of transport as gallons per mile. The cost of transport in squid—at least for the fast-moving ones near the surface—is atrocious, two to four times greater than that of the other pelagic jetters of the same size, and five to twenty times greater than that of pelagic fish. This is why squid use their fins for routine forward motion, and typically save their jet packs for rapid escapes. Given the size and emptiness of the open sea, these large differences in how much food you will need to burn to find your next meal could be critical.

Medusae take a different approach. Medusae come in several flavors, but are all cnidarians, a large group that also includes anemones and corals. One subgroup of medusae is the classic jellyfish that you may have seen washed up on the beach. Another, more diverse group is known as the hydromedusae, which are seldom seen unless you spend time far out to sea. The third group is the box jellies, infamous on Australian beaches for their painful and sometimes fatal stings. What unites these three groups is that they all are jellyfish shaped, meaning that their body is a circular disk that trails tentacles. Some have flat disks, and some have disks shaped like deep bowls, but regardless of the shape of the disk, all the animals move by pulsing it. Even though the disk, known as a bell, can be clear as glass, it nevertheless has muscles inside that can deform it in a way that is much like cupping and straightening your palm over and over. With each pulse, some water is moved backward and the medusa moves in the opposite direction.

As it turns out, it is more efficient to pulse more rapidly when you're small and more slowly when you're large, but all the medusae are different from squid in that they move a large body of water (relative to their body size) slowly instead of shooting out a narrow jet of fast-moving water. This difference has two main consequences. First, medusae are often very slow. Although a few small ones can be quick over short distances, the large ones, such as *Deepstaria* (the size of a jumbo trash bag), move so slowly you wonder if they're even trying to move at all. Second, their cost of transport is low, about equal to the most efficient fish of the same size. This is probably good, since the energy reserves in medusae are minimal. If squid flesh has the calories of a soft drink, a medusa would have the calories of flavored water. I think of medusae as water that has come to life. One of my favorite blue-water dives was far off the coast of Maryland, and we found ourselves at a depth of about 40 feet among a swarm of dinner-plate-sized moon jellies. These animals, usually about 4–6 inches across, wash up on sandy beaches in many parts of the world. The washed-up jellies look like eroded disks of clear rubber, occasionally with a four-leaf-clover arrangement of orange gonads. But in the water, they are deliberate and graceful swimmers. The moon jellies on this dive were all at least a foot across and light pink, and some of the larger ones were surrounded by flotillas of colorful little fish. We were only among them for forty-five minutes, but whenever I want to feel happy about how my life has gone, I think of them.

The third group of jetters comprises the salps. These animals are more closely related to us than any other nonvertebrate animal, but you would never guess it. All of them are transparent, though some have a yellowish tinge, and a few have blue stripes. They're also all short cylinders, usually two to three times as long as they are wide, and open at both ends. They range in size from a pinky nail to a small fist, and are fascinating in a number of ways. First, unlike other jetters, which suck in water and spit it back out on the same side, salps suck in water from the front and spit it out the back. This unidirectional flow is more efficient than other forms of jetting, where the water has to stop and turn around inside the body. They pull off this trick by alternately opening and closing the front and back of their body. The body is ringed by muscles; when they contract, the body becomes narrower. During this contraction, the front opening is shut and the back is open, so the water shoots out the back. When the muscles relax, the elasticity of the body makes it expand to its former shape. During this phase, the back opening is shut and the front valve is open, so water flows in from the front to start the next cycle.

Taking efficiency to an even further extreme, the salp also uses the water flowing through its body to eat. Inside the body is a filter, which catches microscopic algae and other edible particles from the flowing water. So, the body cavity of a salp is not only a jet engine but also a mouth and digestive tract. In fact, the animals are so good at filtering water that when there are large numbers of them (known as a salp bloom), they essentially empty large volumes of the upper ocean. The inedible parts of what the filter catches are compacted into fecal pellets, which drop to the seafloor thousands of feet below. More about this when we discuss food in chapter 6, but I often think of them as fleets of happy little trash compactors.

The icing on the efficiency cake is that most salps travel in chains (figure 6). The individuals in these chains are clones of one another, produced via an asexual budding process. In many cases the chains look like a conga line of salps, but in certain species, the "chain" is more like an organized blob of attached animals. Some are arranged in spirals, and some look

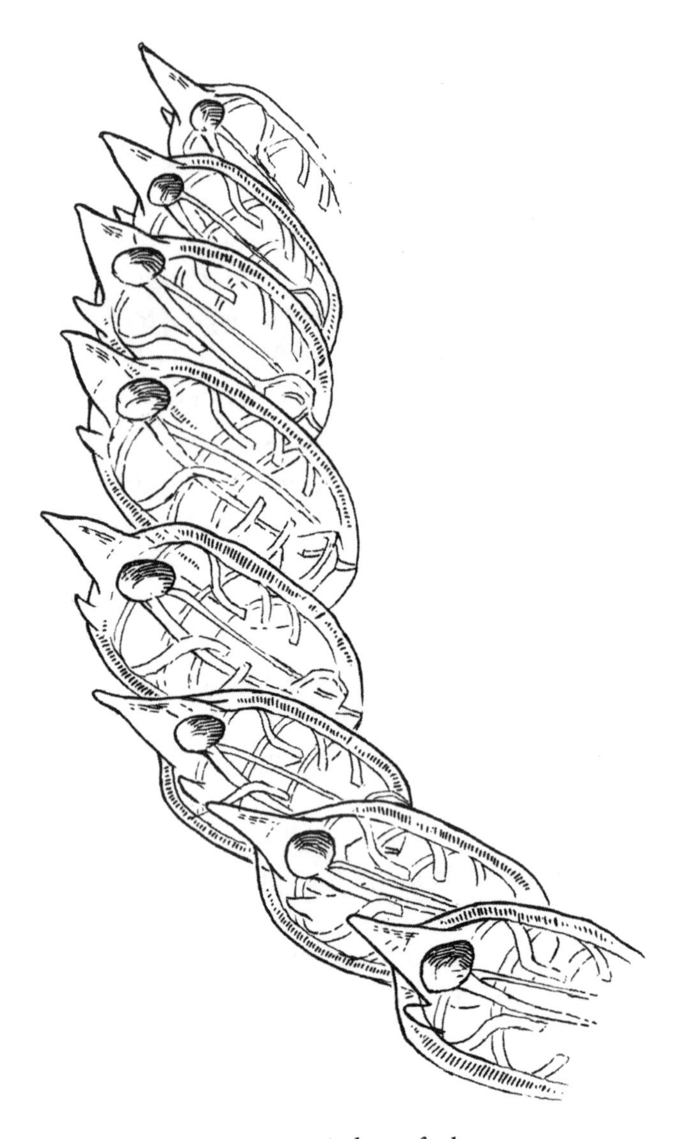

FIGURE 6. A chain of salps

like chandeliers. We do not know all the reasons why salps are often found in chains, but we do know that it allows them to be even more efficient swimmers. You would think that—like a group of friends trying to push your car out of the snow—they would all pulse together. But it turns out that they all pulse to their own rhythm, and that this gives the clutch of clones the smoothest, fastest, and most efficient ride.

Most salp chains I've seen have been only a few feet long, but salps have a closely related group called the doliolids, and I've seen them in chains well over 10–15 feet long. In another "dive for the ages," I was surrounded by over a dozen long doliolid chains, all slowly swirling around me like ethereal serpents. As with salp chains, the members are all clones, but in this case the front individual, called the "nurse," does all the work. The nurse is typically much larger than the other individuals and moves like a salp. I'm not sure why doliolid chains work that way, but it never seemed fair to the nurse.

The fourth group comprises the siphonophores, an extraordinary group of animals if there ever was one. We discussed the best-known example of them, the Portuguese man-of-war, in the buoyancy chapter, but the majority of siphonophores live below the surface. Animals in this group come in an astonishing variety of body plans and an equally astonishing range of sizes. Some are almost indistinguishable from tiny, clear jellyfish, while others can be tens of feet long with so many tentacles and appendages of differing garish colors that it can give you the willies just to look at them. The underwater ones all travel using the same basic mechanism. They use nectophores, which to me always look like a hybrid between a medusa and a salp. They have the deep bodies of salps, but are open only at one end like medusae. Like both salps and medusae, though, they pull in water and then eject it to move. Some small siphonophores have only one or two nectophores that pull along a body that appears to be mostly a tentacle or two, but the larger animals may have one or two dozen of these pulsing bells.

Altogether, the jetters, especially the non-squid ones, are hypnotizing to watch. What surprised me on my first dive was how peaceful the underwater ocean world was. My wife rides horses, and what she loves most about them is that—while they can get scared of a garden hose—they pull you into their naturally peaceful nature and almost require that you join them in an eternal present of sun and grass. I don't think the jellies are requiring anything of me, but I always slow down when I'm around them. Even though I'm a gangly interloper in a rubber suit with a metal tank of air that constrains my visit time like an hourglass, for that time I can pretend to be a jellyfish.

## The Paddlers

There are a large number of oceanic animals that paddle (i.e., push) their way through the ocean. These include crustaceans (shrimp and shrimp-like animals), and animals that are seen by fewer people, such as pelagic snails, pelagic worms, and comb jellies. There is enormous diversity of form among these animals, and of what they use to paddle with.

In some ways, one can think of crustaceans as the insects of the sea. Like insects, they are arthropods, which means they have a stiff external skeleton and many pairs of jointed legs. Also like insects, they are impossibly abundant and diverse. If the terrestrial god had "an inordinate fondness for beetles," as British evolutionary biologist J.B.S. Haldane supposedly said after noting how many beetle species there were, Neptune had a thing for crustaceans. Pelagic crustaceans include the tiny, usually one-eyed copepods, the ostracods (think tiny shrimp swimming inside a tiny clam), and an absolute host of shrimp and shrimp-like animals of all sizes, shapes, and colors. Although some shrimp can move quickly by rapidly curling their tail (the part we eat), this "tail flip" can be done only a few times in a row and is reserved for escaping from predators. Otherwise, all the crustaceans move by paddling some subset of their appendages. In the case of copepods, the appendages are their legs or a pair of oar-like antennae. Ostracods primarily use the middle pair of three pairs of legs on their thorax, extending them outside their spherical shell like some secretive rower. The shrimp and shrimp-like animals typically have paddle-like appendages on the rear half of the body. On average, they move slowly, though I have seen some species, such as krill and amphipods, zip around in large spirals. Even the fastest aren't remotely as speedy as a fish, though. They're of course much smaller, but at the same time the bigger ones tend to be even slower.

The pelagic snails and worms are two groups of animals that make me happy to be a biologist. In addition to being eerily beautiful, they are such a departure from their relatives on land and on the seafloor. From childhood, our image of a snail is an animal so overwhelmed by its shell that it barely moves, and our image of a worm is an animal that leaves the soil only to die on the sidewalk after a heavy rain. But in the pelagic realm, they've both been released to soar like butterflies from a cocoon. Watching them is like watching turtles fly—which I suppose they actually do in the form of sea turtles. Many pelagic snails are called pteropods, which means "winged feet," and the normally flat foot you see in most snails has been split in two and stretched out into wings that allow the snails to fly underwater. The smaller ones are typically heavy for their size and so must flap their wings frantically to keep from dropping, as we discussed in chapter 2, but the larger ones are less dense and move through

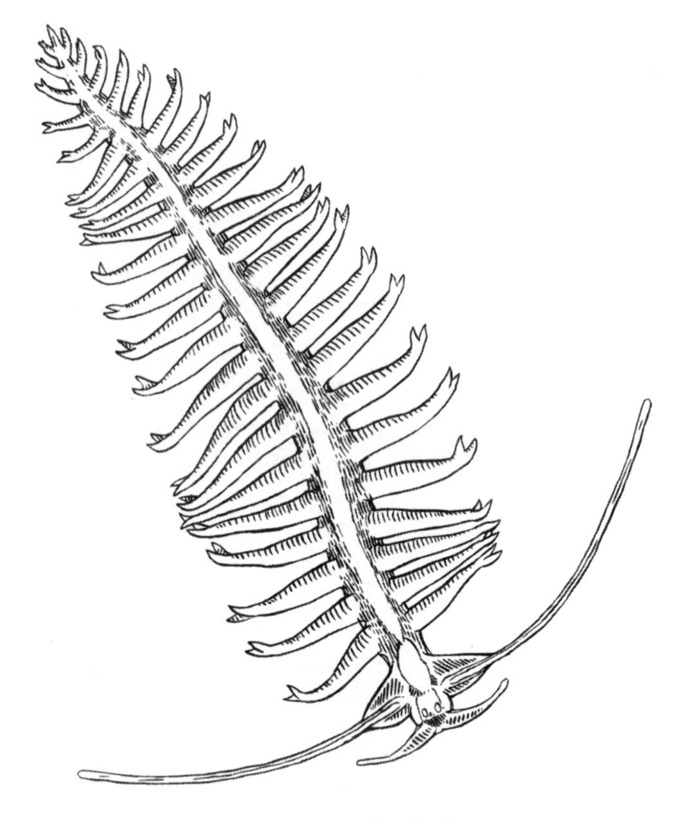

FIGURE 7. A tomopterid polychaete worm

the water in the stately fashion of jellyfish. Many of the pelagic worms also have wings of a sort, but instead of just two they have two per body segment. One of many groups of pelagic worms is the tomopterids, or gossamer worms, which have a dozen or so segments, with twenty-four or more paddles that all work in concert to propel the animal through the water (figure 7). Tomopterids are often transparent, and to me have always looked like crystal chandeliers that have come to twisting life.

Many of the last group of paddlers would also come in last in a race. These are the comb jellies (figure 8), also known as the

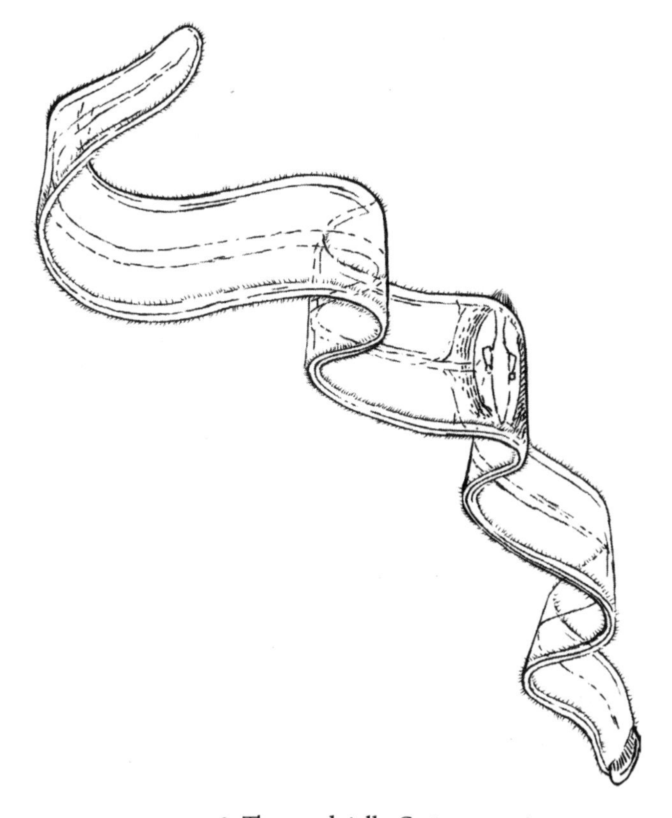

FIGURE 8. The comb jelly *Cestum veneris*

ctenophores (Latin for "comb bearer"). The comb jellies are often superficially similar to medusae in appearing to be made of clear jelly, but they differ in having eight rows of paddles, known as "comb plates." The "teeth" of each comb plate are cilia. Cilia are normally microscopic hairs that are attached to the outside of cells. For example, we hear because there are two tiny snail-shaped organs inside our head known as cochleas. The inside of each cochlea is filled with cilia, which wave back and forth when sound enters the ear. The cilia are each sensitive to different sound frequencies, and so divide our aural world up into notes. The whole system operates much like running a piano or harp in reverse. We also have cilia attached to the cells that line our respiratory tract. These cilia run a nonstop up-escalator that moves mucus, bacteria, and inhaled gunk out of the lungs and into our mouth to be swallowed. Maybe unappealing, but necessary. The cilia on comb jellies are much bigger, and—as the name implies—are bound together in combs. By a lucky optical accident, the cilia are about the same size as the grooves on a CD, so they interact with light in the same way and are gorgeously iridescent. So, in short, a comb jelly is an almost invisible animal wrapped in iridescent bands. As pretty as this all is, though, it is not a recipe for rapid motion, and many comb jellies move as slowly in the water as most snails do on land. Imagine putting twenty toddlers in a rowboat and having them each move just their fingers though the water. Indeed, ciliary locomotion is typically so slow that certain comb jellies have also evolved medusoid locomotion, using two large lobes that flap together like a jellyfish. But even some of those that only use comb rows can be zippy for their size. I suppose even comb jellies can become impatient.

## The Drifters

My former graduate student Julia often says, "Fish don't know they're in water." For her this has a political meaning, but it also makes me think about how fish—and all other oceanic animals—don't know they're in a current. The ocean is awash in currents at all depths. The surface currents, such as the Gulf Stream that runs from the Gulf of Mexico to Northern Europe, are caused by prevailing winds interacting with the rotation of the earth. The end result of these forces is clockwise loops in the North Pacific and the North Atlantic and counterclockwise loops in the South Pacific, South Atlantic, and the Indian Ocean, with additional currents in both polar regions. The strongest currents are on the eastern side of continents in the northern hemisphere, where they bring warm water to colder regions, and on the western side of continents in the southern hemisphere, where they bring cold water to warmer regions. An example of the first is the aforementioned Gulf Stream, which moves at about 4–5 miles per hour. An example of the second is the Peru Current (also known as the Humboldt Current), which travels the western boundary of South America at speeds of about 10–12 miles per hour. The strongest current of all is the Antarctic Circumpolar Current, which screams around Antarctica at 15–25 miles per hour.

These are just the surface currents. Deeper down there are other currents, and many of them go in the opposite direction to the surface current directly above them. For example, deep beneath the Gulf Stream is a current that takes water back south. The deep currents, which are caused by changes in density due to changes in the temperature and salinity of the water, take an enormous amount of water on a thousand-year-long convoluted journey through many of the deep places of the ocean, in what is known as the "global conveyer belt." Together, the surface and deep currents have enormous implications for our climate and weather, and for which parts of the ocean have enough nutrients to support abundant life. Here we focus on what the water movement directly does to animals.

As I mentioned before, humans are pitiful swimmers, on par with their dogs. The quickest of us can manage 5 miles per hour for a minute, but most of us are lucky to swim 1–2 miles per hour, and not for long. This is well under the speed of the major surface currents, and also why we are so vulnerable to riptides. As pitiful as we are, though, the bulk of ocean animals are even more so. Fish, sharks, squid, sea turtles, and most marine mammals can make headway against ocean currents and thus choose their fate, but everyone else is at the whim of a force that will at best forever cycle them round and round half an ocean basin. At worst, the current will drive them into the polar regions to freeze or into the still centers of these circles to languish.

This "everyone else" includes all the non-squid invertebrates—which means the shrimp-like creatures, the various jellyfish and other gelatinous animals, the pelagic snails, the pelagic worms, and a host of other enigmatic animals. It also includes the pelagic larvae of many animals, which we will discuss in chapter 7. Life for these drifters is determined by the currents that carry them. As I mentioned above, though, these animals may be completely unaware of this. This is not as strange as it sounds. After all, we have no personal awareness that the earth is moving at 67,000 miles an hour around the sun. Similarly, hot-air balloonists at night may not know that there is any wind at all, let alone one that is sending them rapidly across Kansas. The same is even more true in the open ocean, where reference points such as the ground are not available. About fifteen years ago, we did a blue-water night dive off the Bahamian island of Little San Salvador. As with other blue-water dives, we were tied to each other and to the boat, all of which drifted with the current at the same speed, so we had no sensation of movement. The current was unusually strong that night, though, and we dove for a long time, so when we surfaced and asked the boat driver how far we'd drifted he told us it had been miles, and there was a different island in view. As always, knowing something is not the same as living it—it was disturbing to move so far (especially at night) without any sense of motion at all.

The drifters don't move just a few miles but potentially thousands, depending on how long they live. But to them it's as if they're not moving at all, since the same large chunk of water is traveling with them. However, some may wish to have more control over their fate. For example, they may wish to avoid moving into waters that are too hot or too cold. Or they may wish to avoid the coasts, or to move toward the coasts so that they can start the next stage of their life. One solution, of course, is for a species of drifter to be extremely numerous, so that even with no navigational control some animals will get to where they need to go. But certain drifters seem to do better than others, and researchers have proposed that they do this not by fighting the currents but by changing their depth.

Earlier we discussed that surface currents often have currents underneath them, and that many of these currents go in the opposite direction to the surface current. So, in theory, if you can control your depth and have some understanding of the currents, you can travel the earth simply by going up and down. As with many good ideas, it's nearly impossible to determine who came up with it first. The oldest mention that I know of comes from Alister Hardy, a British plankton biologist who worked mostly in the first half of the twentieth century. His mind went in many directions, but he is a hero of mine because he combined science with both good writing and excellent art. He painted many of the animals he saw at sea (at a time before color photography was common), making watercolor montages that I often use in my talks. As a scientist, he developed the Continuous Plankton Recorder, which was one of the first tools used to determine the distribution of the ocean drifters. He suggested that vertical migration—a daily up-and-down movement that we'll discuss more later in this chapter—may be a way for plankton to stay in place. During the night, the animals would be pushed in one direction by the surface current, and at daybreak they would go deep and be pushed back to where they started by a deep countercurrent. By changing the amount of time they're in deep and shallow currents, drifters could have some control over their location. Also, since many animals can change depth by changing their buoyancy, this form of travel could in theory require little effort. Of course, the devil is in the details. Not every current has a countercurrent underneath it, and—as mentioned above—deep currents are often much slower than surface currents. Also, to take advantage of these differing currents, one must know both where the currents are and where one is in relation to them.

Tom and Donna Wolcott at North Carolina State University have approached this concept of depth-controlled drifting by building what they term "plankton mimics." They do not look remotely like the drifters they mimic but are instead drifting robots that can determine where they are via GPS and also control their depth. I've followed the history of their project and visited the lab where they were building the mimics. Like all experimental oceanic work, the project had many challenges and setbacks. As we like to say, "Don't put anything in the ocean that you ever want to see again," and I would guess that many of the mimics are still floating somewhere far out to sea, and will be a confusing object to anyone who finds one. The Wolcotts were eventually able to deploy a set of them off the coast of Northern California and show that the mimics could roughly maintain their position in nutrient-dense waters by changing their depth at the right times. There is much that remains to be learned about the drifters and whether they are in fact much more intentional in their movements than they first appear, but the Wolcotts' work shows that is at least possible.

## Vertical Migration: The Night Visitors

As we discussed in previous chapters, the open ocean varies much more rapidly in the up–down direction than it does side to side. One can travel for tens (in some cases hundreds) of miles horizontally in the ocean and see nothing but more of the same, but if you sink just 200 feet, the light level will drop at least tenfold, making it harder to see predators, prey, or even animals of your own species. Ultraviolet radiation drops even faster; for example, staying 60 feet down or going to the surface can mean the difference between an uneventful day or a brutal sunburn. Also, if the surface waters are not mixed by wind, the temperature will drop like a stone as you descend.

So, it's unsurprising that animal communities in the water column are primarily structured by depth. Some species are always found at the surface, some only in the black depths, and some live in between in eternal blue twilight. But many animals move between these depths on a daily basis, performing what is known as diel (daily) vertical migration. And by "many" I truly mean *many*. Diel vertical migration is by far the largest migration on earth in terms of both mass and the number of animals that go up and down. According to Kelly Benoit-Bird at the Monterey Bay Aquarium Research Institute, the global mass of the small migrating fish alone is about a billion tons. Given how little these fish weigh, we're talking about trillions of fish. There are also uncounted numbers of krill, copepods, and other crustaceans, along with hordes of cephalopods, salps, and other animals.

There are variations, but the typical pattern is for vertical migrators to stay deep during the day, start moving up to the surface near sunset, spend some of the night at the surface, and get back down before daybreak. The distances they travel are substantial, especially when one factors in how small most of these animals are. Usually, the change in depth is around 500–1000 feet, though some salps swim up 2,500 feet every evening. Even a human might find swimming 1,000 feet up and 1,000 feet down every night to be a major addition to their exercise regime, but many of these animals are less than an inch long. For certain krill (which are about an inch long), it's been calculated that their daily migration is twenty to thirty thousand body lengths long and is done over a period of about six hours (three hours going up, three hours going down). For a 6-foot human, this is running a vertical marathon every night, except of course you have to swim it.

In sum, vertical migration is a challenging daily task performed by trillions of animals from a wide diversity of species, which has made it a central area of study ever since it was discovered by World War II-era sonar operators. Any time biologists see so many organisms doing something that hard, the two first questions are *why* and *how*. Both questions have been complicated by the fact that vertical migration also occurs in lakes. It is easier to study in lakes; because the migration distances are on the order of only 10–20 feet, so much of our understanding of this process comes from freshwater systems. But, in at least some ways, the *how* and *why* answers for freshwater migration seem to be different from those for marine migration. As always, we need to be careful.

Although other explanations have been put forward (such as energy conservation and temperature preferences), it's generally accepted now that the primary answer to the *why* question is predation. We'll discuss this in more detail in the next chapter, but predation in the water column of the open ocean is intense and has led to the evolution of some equally intense responses. The argument for diel vertical migration is that it allows animals to hide from predators in the darker depths during the day and come up to the surface at night to feed. This argument makes intuitive sense, since a drop of 1,000 feet in clear water will decrease the light levels one to ten-thousand-fold, which is about the same as going from a sunny day to mid-twilight. You might be thinking, "I can see someone just fine during mid-twilight," but remember that many of these animals are tiny, and that even clear water is harder to see through than air. I can say from experience that it does get a lot harder to see a small object when you're at 1,000 feet, unless the submersible pilot turns the lights on. However, we are assuming that most oceanic animals hunt by sight, which of course is not true for some predators, such as whales and dolphins, which use sonar, and others may use smell or touch.

There is direct evidence for the predation hypothesis, but due to the challenges involved in working with oceanic migrators, most of it comes from work done with freshwater migrators in large outdoor enclosures known as "mesocosms." For example, we know that certain freshwater plankton will stop migrating if you take predatory fish out of their habitat. The story for freshwater vertical migration is more complex, because certain migrators in lakes will sink to lower depths during the day in the presence of normal levels of UV radiation. This is unlikely to be an issue for ocean migrators, because they travel far deeper than UV radiation ever penetrates.

Of course, the predation hypothesis explains only why they stay down, not why they come up. After all, they could stay down all the time and skip the exercise. The surface waters do have a couple of advantages, though. One is that they are warmer, and having a warmer body allows you to move faster, see better, and in general run your various bodily processes at a faster rate. A second is that the twilight light levels that many of these animals experience at depth during the day drop to pitch black at night, while the surface waters will have some light if the moon is out. It's easy to miss if you live in a city, but the full moon is terribly bright, making the night world hundreds to thousands of times brighter than it would be without it. For much of the winter, the full moon shines directly into our rural bedroom, and it's so bright that we often think a helicopter must be flying overhead. In fact, sometimes the moonlight over the ocean is so intense that it can push the animals back down a bit, especially in the endless night of the polar winter.

The third reason for going up is, frankly, that everyone else is doing it. Because so many animals vertically migrate, the density of animals near the surface skyrockets at night. In a way, vertical migration is like an enormous compactor, taking a large fraction of the animals in the top 2,000 feet and squeezing them into a band at the surface that is maybe only 200 feet thick. So, you can skip the party and stay down at night, but the prey levels around you will drop as everyone else leaves. If you do go up, though, prey levels may be much higher than you experience during the day. The standard statement in marine biology classes is that vertical migrators stay down during the day to avoid predation and come up at night to feed, but this doesn't really convey the creepiness of it. Imagine a world where zombies rule your city during the day, forcing the humans to hide in their houses, getting hungrier by the hour. Night falls, the zombies go to sleep, and everyone runs to the supermarket. Except instead of buying the food, they spend the night in the store eating one another. Now add the fact that some people have been waiting in the store all day for the people to show up at sunset, so that they can eat them, and you get an idea of the situation. Some oceanic animals, however, such as krill and salps, rise to graze on only phytoplankton; I suppose they would be analogous to the humans in the supermarket quietly eating the vegetables.

So that's the accepted answer to the "why" of diel vertical migration. The other big question is how the animals know when it is time to go up. Although there are some animals that go up and down twice in a night, and others that "reverse migrate" by going down at night and coming up during the day, the majority start moving up at sunset. Since we also know that a full moon can encourage the animals to go back down, and that eclipses can trick them into coming back up, if only for a little while, the strong suspicion is that vertical migration is regulated by light levels. There are multiple hypotheses about how light levels trigger upward migration at sunset.

The first is known as the isolume hypothesis, which suggests that animals, much like Goldilocks, have a narrow range of preference for light levels. So, as the light levels drop at sunset, the animals move up to stay in waters of the same light intensity. The other two primary hypotheses are based on the special things that happen at sunset. One is of course that it gets dark. So, a trigger could be the light dropping below a certain level; this is known as the threshold hypothesis. The other is that light levels are changing quickly. It's hard for us to appreciate how fast light levels drop after sunset. Night falls quickest in the tropics, because the sun goes straight up and down. In these regions, light levels during twilight drop a hundred-thousand-fold fold in just thirty minutes, or tenfold every six minutes. At more northerly or southerly latitudes—say, 45 degrees from the equator—the sun sets at more of an angle and twilight is a bit longer, but still only about forty-five minutes. This rapid and sustained drop is much more intense than any thunderstorm or eclipse and thus could in theory also be used to tell the animals hiding in the deep that it's time to go up.

So how do we find out which hypothesis is true? This has been done in labs where researchers put the animals in enormous, water-filled mesocosms and manipulate the light levels to see what changes encourage them to move up. It's tricky for many oceanic animals, though, both because they're hard to keep alive and because they often migrate over hundreds to thousands of feet. So, people have also attempted field experiments. The first cruise, which I have already written about, was an attempt to do exactly this by my advisor Edie Widder and her then postdoc Tammy Frank. Like many oceanic field experiments, it was breathtaking in its combination of high and low tech. The high-tech aspects were the use of the *Johnson-Sea-Link* submersible to follow the animals, and the development and use of a contraption called LoLAR, which could measure the extremely low light levels in the deep. LoLAR, like just about every piece of oceanic instrumentation, is a heavy metal cylinder with a cute name. In this case, LoLAR stood for "low-light auto-calibrating radiometer" because it continually calibrated itself with a tiny glowing sphere powered by tritium gas—the same gas that is found in nuclear warheads. The low-tech aspect was that the animals were counted during the dive by Tammy (who was in the front) frantically yelling out every one she saw through the large bubble window, and Edie (who was crammed in the back) writing down what she heard on a pad of paper with equally frantic haste. On later cruises for the same project, I got to be the recorder in the back, and if you ever fell behind in the count you were so doomed. I can still hear "squid squid fish squid jelly squid jelly jelly!" in my head.

By monitoring both the light levels and the depths of the animals it was possible—when combined with similar work by others—to answer the "how" question. The answer, which is so common in biology and can be either uplifting or dispiriting depending on your world view, was that different animals did different things. Some seemed to follow an isolume, while others appeared to care more about a threshold light level or the rate of change of the light levels during twilight. If there is one thing I have learned in my over thirty years as a professional biologist, it's that if there's a way to do something, some animal is already doing it. Although mostly driven by evolution rather than what we would call intention, the innovation and resourcefulness of the natural world are astonishing. So far, we've seen this resourcefulness in how pelagic animals solve the major physical problems of the ocean—gravity, pressure, light, and getting around the enormous habitat. But life is more than just surviving physical challenges; you also need something worth living for. So, next we'll explore the lives of these animals.

# CHAPTER 6

# Food

> If you don't hunt it down and kill it, it will hunt you down and kill you.
>
> —Flannery O'Connor, *Wise Blood*

## Our Shrinking Larder

Research vessels are homes, transports, small factories, and floating scientific labs. But perhaps most important to us is that they are massive refrigerators. This is not because their interiors tend to be kept cold (they do—for reasons I've never understood) but because they hold a tremendous amount of food. The front-most triangular section of the main deck, in front of the galley and the mess, where the food is cooked and eaten, respectively, can contain enough in refrigerators, pantries, and freezers to feed a crew of thirty for months. In addition, there are sometimes refrigerator-sized freezers on the main deck filled with pint containers of ice cream—what one crew member used to call the "single-serving size." Many ships I've been on also have hollow benches in the mess, with hundreds of candy bars and single-serving packs of chips and microwave popcorn under each seat cushion. It's a snacker's delight, and as we work into the early morning hours, we tend to raid the food bins, because—as my former student Nick Brandley used to say—"food equals sleep." You can stave off sleepiness and boredom for another forty-five minutes by downing another bag of tortilla chips. In the first chapter, I said that the ocean doesn't have a smell, but the ships certainly do. Outside, they smell of diesel, and inside they smell of ice cream mixed with the "butter" in microwave popcorn.

Fresh food is another story. In the days before each cruise, the cook and steward do the mother of all shopping runs at whatever happens to be the local supermarket. They buy multiple shopping carts of produce, meat, and other fresh foods. It's impressive to watch them crane it all on board, but it doesn't last nearly as long as you'd think or hope. Over time, the foods are eaten in order of perishability. I was on a cruise where salad in the first week consisted of lettuce. Then we moved to spinach, and from there to raw cabbage, and finally chopped-up raw brussels sprouts in week three. The cooks vary in their ability to make the food seem fresh longer. One cruise started with creamed beef hash (not what you want to see on those early seasick days) and went down from there, but another was a two-week Cajun feast. We all tell stories of the cruises that had either amazing or terrible food. My friend Tammy Frank once had a "raw cruise" where none of the meat was fully cooked, and I still get queasy thinking about that first "Nutra-Fat" cruise. We obsess about food, because 90 percent of the time life at sea is crushingly dull, with only meals and sleep to mark the passage of time. Sandra Brooke, whom I sailed with many times, said that life at sea was like being a domestic animal, always waiting for the next meal.

Fresh or not, the meals and snacks on oceanic ships are likely the greatest concentration of calories for many miles in all directions. The open ocean is on average quite empty, and what is there is often nutrient-poor. There are some large, dense schools of fish or crustaceans, but these are usually found either closer to the coast or in polar waters. In the warmer, offshore waters that I usually work in, there's generally not much down there. On many of our cruises, we fish with a net that has an opening that is 100 square feet in area. We usually tow the net for hours at 1 mile an hour, and we often do it right after sunset to catch as many of the vertically migrating animals as possible. So, in three hours, we would have pulled the net through 1,500,000 cubic feet of water, the same volume as the dome of the US Capitol (which is 100 feet in diameter and 200 feet tall). After doing all this through waters that should be dense with animals owing to vertical migration, we might catch only a small bucket of animals. If we towed deep during the day, we might catch almost nothing, what Edie once called "straining water." It is likely that climate change, by warming the ocean, has made it emptier; warm water has less oxygen and thus fewer animals. We're also not catching everything we pass, since faster and smarter animals will dodge the net. As the marine biologist Peter Herring said, we catch "only the slow, the stupid, the greedy, and the indestructible."[1] All that said, if you think of the open ocean as an even more gigantic refrigerator than our ships, there is no escaping that it is on average a sparsely provisioned one, with far fewer calories per volume than a coral reef, salt marsh, grassland, forest, or many other habitats.

The relative dearth of food and several other factors change the nature of predation in the open ocean relative to land. First, because the open ocean has occasional food oases surrounded by food deserts, and because these oases move and are thus unpredictable, predation can be more intense. When I give

1. Peter Herring, *The Biology of the Deep Ocean* (Oxford: Oxford University Press, 2002), 21.

talks on this, I remind the audience of the old nature shows where a lion pride sits on a small knoll overlooking a herd of antelopes. Now and then, the lions come down off the hill and kill an antelope or two, focusing (at least as the shows always described it) on the old, the young, and the sick. What you never saw, though, was the lion pride coming off that hill and devouring the entire herd—because it doesn't happen. But you do see this in the open ocean, where a group of large fish will come upon an enormous school of small fish and—over the course of several days—kill and eat the whole school. There is astonishing footage of this in the BBC's *Blue Planet*.

On a more personal note, the lack of food in the open ocean relative to a coral reef has impacted my interactions with sharks. I've seen many sharks on coral reefs and never paid much attention to them. I tell my students not to worry about sharks on reefs, because "we're swimming inside their refrigerator." After all, would you eat the odd-looking rubbery thing at the back of the fridge that you'd never seen before, when all your favorite foods were right there? On a blue-water dive, though, if a shark shows up, you and your dive buddies might be the first promising objects it's seen in a while. The sharks on these dives are always "interested" and occasionally have done mock attack runs at us. We've never felt any real danger, but have taken this as our cue to leave.

The open ocean also differs from most terrestrial and coastal habitats in that prey animals are more vulnerable. First, predation is three-dimensional, in that you can be attacked from any or all directions at once. Some birds are attacked in the air, and various small ground animals are preyed upon by overhead birds, but in general most animals in other habitats can safely assume that they will not be attacked from above, below, and the side. In certain cases, a pelagic animal can be attacked from above and below at the same time, and by different species. These scenarios have a way of wheedling themselves into the darker corners of your mind, and I know a number of bluewater scuba divers who have nightmares about being first stalked from below and then eaten.

The third and perhaps the biggest problem about oceanic predation is that you can't do anything about it. The main issue is that there's nowhere to go. If a rabbit is being chased on land, it can use its impressive maneuverability and short-term speed to get to its burrow or another hiding place before it tires. But in the open ocean, there's never a safe spot. In addition—unlike on land where an adult buffalo would never eat a baby lion—a larger fish or squid is more than happy to eat anything smaller, which means that any animal larger than you might eat you. And because of how locomotion works underwater, the larger animal is nearly always faster over any relevant distance. So, there's nowhere to go, and in general you can't outrun your predator. Finally, because the open ocean is featureless, hiding is hard. You need to look like water.

Together, the difficulties of finding food have led to some novel feeding strategies, and the difficulties of avoiding becoming food have led to a number of special camouflage strategies. I often give talks entitled "Hide and Seek in the Open Sea," and it's just that. Beneath the surface, an ancient, complex, and very serious game is going on, which we'll explore in this chapter.

## Fishing with a Net

Many pelagic animals hunt their prey and eat them, much like you'd imagine a lion would. We've already discussed the schools of pitiful fish being devoured by sharks, fish, and birds, and there is a host of other species that follow the terrestrial example of chasing down an animal and biting it until it dies. Some, such as sperm whales and pods of orcas, are truly formidable and often considered the most impressive predators on the planet. Schools of large squid can also be highly successful hunters, and Humboldt squid in particular are daunting. Although not as long as giant squid or colossal squid—instead, about human-sized—they make up for it by being meaty, having sharp beaks, lining their suckers with "teeth," and hunting in packs. We do not know whether they hunt in coordinated packs as lions and wolves do, but recent work has shown that they have a complex repertoire of about fifty color patterns on their bodies, which they can cycle through. For all we know, most of these signals could simply be transmitting, "My fish, my fish!" to the other squid, but why have fifty words for that? I personally would be both thrilled and nervous to be surrounded by a pack of human-sized, tentacled predators sending colored signals to one another in a twilit sea.

But I'd like to focus here on the animals that do things differently from those on land. As I mentioned, the open ocean, especially in warmer waters, can be quite empty. In addition, what is there can be quite small. Imagine a world where the only food is the dust that sparkles in the morning sunbeams of your house. This dust is in fact nutritious, some of it being flakes of your own dead skin, but you'd go crazy trying to pick it out of the air with your hands or mouth. What you could do, though, is get a large square of fine window screen and walk through your house, holding it in front of you. You'd have to be careful about the mesh. If it's too coarse, the skin flakes would go through. If it's too fine, not enough air will go through, and the skin flakes will bounce away. But if you get it right, you eventually get a nice coating of dust on your screen that you could then lick off.

This is essentially what many pelagic animals do. They create a screen and use it to trap tiny particles out of the water. The animals on the seafloor, such as the relatives of starfish called "feather stars" and a delightful group of animals known as "feather duster worms," do this as well and often have the advantage of being stationary in a current. This would be like putting a fan in front of your screen, so that you could just sit on the sofa and wait for your coating of dust to be large enough for a meal. The animals in the water column, however, can't do that, because if there is a current, they are moving with it. They seem to have settled on three solutions.

The first solution is one that we discussed in the last chapter, and that is to make moving and eating the same thing. The salps and doliolids we discussed both have internal filters and collect particles in them by swimming through the water. The second solution is to sink rather than swim, which is done by the pseudothecosomes we discussed before. My favorite one to see on dives is called *Corolla* (figure 9). Including their two large wings, they're seldom bigger than a child's hand, but they secrete a 6-foot-wide diaphanous sheet of mucus and then hang below it. Scaled to human size, the sheet would be about 150 feet across. This mucus net is delicate but sticky enough to catch and trap particles from the surrounding water. To create the required movement, the animal slowly sinks, pulling the web down through the water with it like a sticky parachute. When enough particles have been caught in the web, the animal ingests the whole thing, sorts out the good parts, and then recycles the mucus to make another net.

The final trick is to remain stationary but to be your own fan and move water through the filter. The larvaceans do this, using one of the most remarkable objects in the entire animal kingdom. Larvaceans are closely related to salps and doliolids,

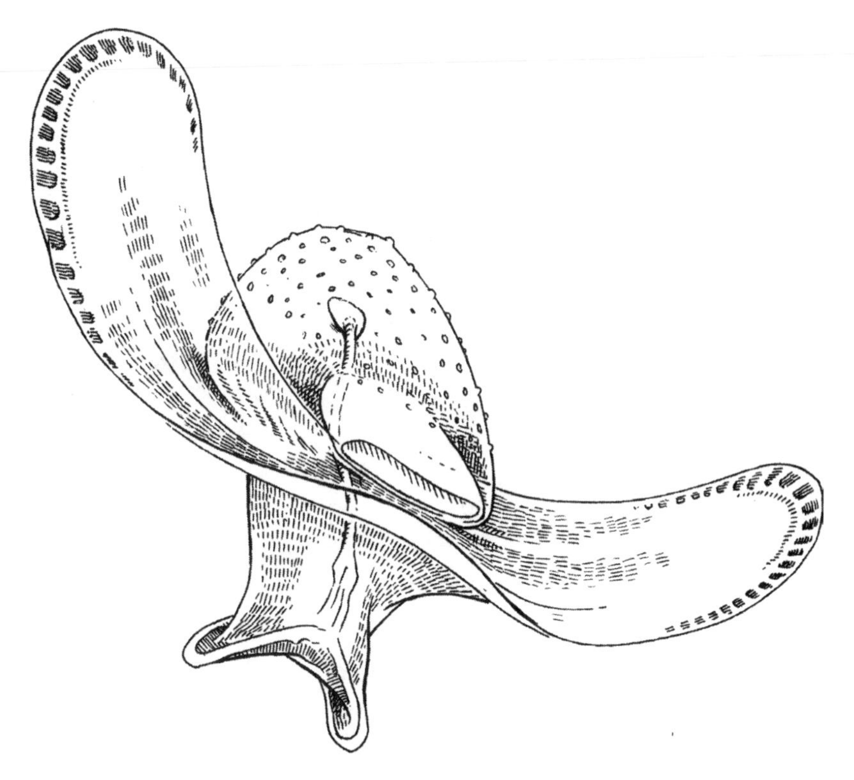

FIGURE 9. The pseudo-shelled pteropod *Corolla*

which means they are also closely related to us as vertebrates. In fact, they even have a nerve cord in their little tail that is considered to be evolutionarily related to our own vertebral column. They look like transparent tadpoles and are at most 4 inches long but are best known among pelagic biologists for the mucus houses they build (figure 10). The "outer house" is a nondescript blob whose purpose (as far as we know) is to screen out large particles. In human terms, it's like the prefilter we use on many water filtration systems. The "inner house"—which is, naturally, inside the outer house—is a true work of art, a complex filtering structure with paired filter webs, a place for the

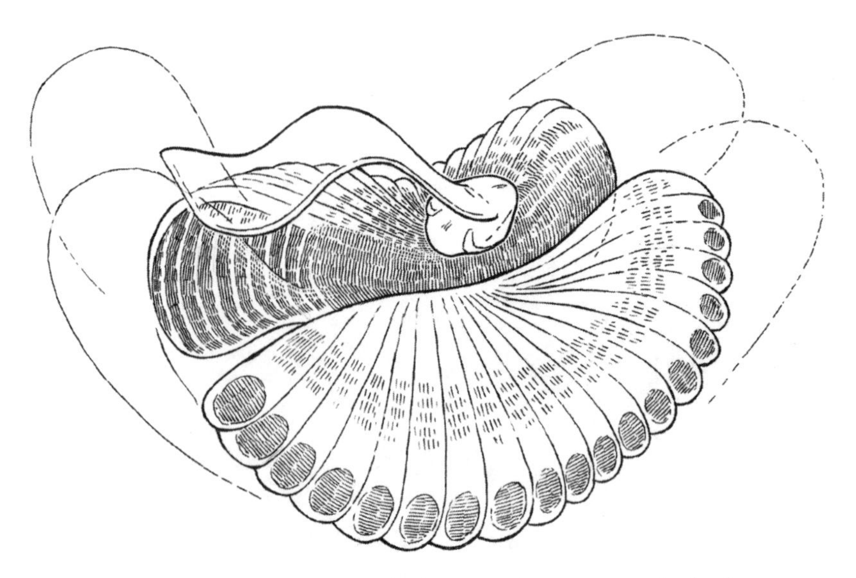

FIGURE 10. A larvacean inside the mucus house it uses for feeding

larvacean to sit and pump water through them both, and even a special exit in case the animal feels threatened. Scaled to human size, the outer house is about the size of a two-bedroom house, and the inner house is about the size of a large bathroom or small bedroom. With the outer house screening out the larger particles, the inner house concentrates the remaining smaller particles, often referred to as nanoplankton. The larvacean eats this concentrated material, further filtering and concentrating it.

The larvacean keeps using its two houses until the filters clog or until the animal dies. Many larvaceans, especially the smaller ones, live only a few days, so the latter may happen before the former, but if the filter does clog before it dies, it will abandon the house and make a new one. These abandoned houses lose their complex beauty and slowly sink, with pelagic biologists typically referring to them as "sinkers." I was on a cruise led by Kakani Katija of the Monterey Bay Aquarium Research Institute where she had a laser scanning system on the robotic sub that was used to make three-dimensional reconstructions of the larvacean houses. It was everyone's job to find them, which meant sitting in the ROV control room on the ship for hours and days with the submersible pilot and watching the monitors. It turns out that sinkers are painfully common, at least off the coast of California, and there was a steady disappointed refrain of "sinker" every time we drove up to a promising blob of mucus only to find it was abandoned.

All in all, the story of filter feeding in the open ocean is the story of mucus. When I started my postdoc at Woods Hole Oceanographic Institution, I had a talk with my new advisor, Larry Madin, about pelagic animals. I wanted to talk about their transparency, which had gotten me excited at that time, but he wanted to talk about "the truly astonishing things these animals do with mucus." I didn't see his point at that time, but after many cruises and dives I sure do now. Oceanic animals have turned a material that we consider disgusting at best into the finest weaving material the world has seen.

There are also oceanic filter feeders that don't use mucus, instead creating systems built from tentacles that work like living spiderwebs. The best known of these are the various medusae—the jellyfish. As we discussed, they come in three major varieties, which are the Hydromedusae, the Scyphomedusae (what we usually think of as jellyfish), and the Cubomedusae (the painful box jellies from Australia and the Caribbean). Although these animals come in many different sizes, shapes, and colors, they all use nematocysts—tiny sacs inside specialized cells on the surfaces of the tentacles, which have poison-tipped harpoons coiled inside them. If the tentacle touches anything, osmosis floods the sac with water, which pops the harpoon out. The toxins are a nasty mix of substances that cause pain and in certain cases affect the nervous system and/or the heart. I've been stung countless times while diving, swimming, or stupidly sticking my hand in a bucket of animals. The stings have ranged from mildly annoying twinges to ropelike lacerations that burned for days. Humans are fairly protected by our outer layer of dead skin; fish and marine invertebrates are more vulnerable, and animals that use nematocysts are highly successful at trapping and killing prey.

The tentacles are deployed in many different ways, probably to maximize the chances of catching certain animals and minimize the chances of catching others. My favorite tentacle fishers are the siphonophores, because some of them use lures. The siphonophore *Agalma*, for example, has a large transparent body. Many tentacles, which are also transparent, extend from this body, but the end of each tentacle has little blobs that look to all the world like larval fish or the little crustaceans known as "copepods." So, what an animal approaching *Agalma* sees is not the main bulk of the animal itself, but a cloud of snacks. If it bites at these snacks, though, it will trigger the nematocysts, harming itself. If it is small enough, it will be too hurt to escape and will be pulled in by the rest of the tentacles to be digested by other parts of the body of the *Agalma*. This particular siphonophore is a weak swimmer, so this strategy—termed "aggressive mimicry"—is a clever workaround.

## Specialists

Given the empty nature of much of the oceanic water column, it's perhaps not surprising that certain animals have evolved to eat ever-smaller things, since smaller things tend to be more abundant. Particularly fascinating are the larvae of eels. Eels, like so many things in life, are something everyone has heard of, but no one understands. This is because eels at different life stages look so different and live in such different habitats. In fact, it took until the late nineteenth century for us to realize that the juveniles and the adults were the same animal. The adults are relatively large, opaque, snakelike fish that live in freshwater streams and have been a food source in many countries for centuries. Their common names typically reflect the regions where they live their adult lives, so we have the Japanese eel, the European eel, the South African eel, and so on. But for the longest time no one had ever seen a young eel, or at least they didn't realize that they had. It turns out that young eels were open-ocean animals that scientists had been calling "leptocephali." The larval eels look nothing like the adults, looking instead like long transparent leaves. The larva are extremely flat, with tiny heads, and seem to have no internal organs, though you can make out a digestive tract. The details vary among species, but in general the story is this. The adult freshwater eels travel a great distance across one of the ocean basins to spawn, with the young being these leptocephali. These travel the open ocean for a few years, at some point returning to the coast. There they transform into their young adult "glass-eel" stage, which is more eel-like in shape and often both silvery and transparent. These glass eels then make their way back to freshwater, often crawling across wet grass, and back in freshwater they transform into adult eels. It's a story that we're still trying to sort out and has led to at least one book.

The questions here, though, are what do leptocephalus larvae eat and how do they catch it. As I mentioned, they have tiny heads. These come with even tinier mouths. The animals are also slow, so catching live prey might be difficult. A further mystery is that their guts always seem to be empty. This stymied pelagic biologists, because—since they almost never get to see pelagic animals eat another one in the wild—the usual way to figure out what the animals eat is to cut open their stomachs and look. I always thought this was a grim business, much like an autopsy, but some colleagues of mine appear to relish it. My friend Tracey Sutton at Nova Southeastern University in Florida, for example, spent his entire time in graduate school cutting open the stomachs of dead deep-sea fish and identifying all the half-digested body parts. If you open the stomach of a leptocephalus larva, though, you don't find anything. This has led some researchers to suspect that the animals absorb nutrients through their skin. This is not as crazy as it sounds. There are nutrients dissolved in ocean water, and leptocephali have a lot of surface area for their weight, owing to their flat, leaflike shape. Leptocephali also have very low metabolic rates, but we know that part of the reason for this is that they're storing most of what they ingest in the form of sugars called glycosaminoglycans. They later use this stored energy to fuel their metamorphosis into glass eels. Therefore, they might need more food than you would expect from their sluggish activity.

The third cruise I ever went on was to the eastern Gulf of Mexico, with the goal of figuring out what these animals ate. It was led by Jose Torres at the University of Southern Florida, but also included Tracey Sutton, who taught me as much about oceanic fish as I could hope for (and more about *The Simpsons* than I needed to know). The work from this cruise, combined with later work by other scientists, has so far not told us whether these animals are eating with their skin, but it has shown us one thing that they are eating—marine snow.

Marine snow is one of those things that you never hear or care about, unless you want to see anything underwater in the dark, at which point you care a lot. Marine snow is a charming term for what is actually bacteria-laden detritus. The upper layers of the ocean are packed with tiny forms of life; in particular, uncounted numbers of single-celled algae. After death, these algae clump together with small bits of mucus shed from other animals, and other remnants of life, to create bits of fluff that are the size and shape of snowflakes. Marine snow slowly sinks to the bottom, often taking weeks to get there. The amount of carbon in each flake is small, but marine snow is so common that the total amount of carbon moved from the ocean surface to the deep-sea floor is significant. Under normal illumination, marine snow is so transparent that it's hard to see, but if you shine the floodlight of a submersible or camera at it, enough light reflects back to make every single video and photo you take look like it was taken in a snowstorm. It makes finding anything in the ocean that much harder, and I develop squinty eyes after a couple of weeks of staring at the camera feeds of robot submersibles trying to guess whether I see something interesting or just more snow. This odd fluff turns out to be a food source. Not a highly nutritious one, but one that both is common and can't run away. Leptocephali appear to take advantage of it. Likely, other animals do as well, and certainly some of it gets caught by various filter-feeding animals, but this is tough to prove because marine snow is hard to identify in a stomach.

A different group of specialists is the "naked" pteropods. We've discussed pteropods before; they are snails that have evolved to swim through water. They come in three varieties: (1) the "false-shelled" Pseudothecosomata, like the *Corolla* we discussed; (2) the shelled Thecosomata, which look like small land snails with flapping feet; and (3) the shell-less Gymnosomata. The gymnosomes (which literally means "naked body") look less like snails and more like slugs. Like the pteropods in the other two groups, they have wings, but those are often disproportionately small. However, the animals appear to be more muscular, since all the ones I've seen are agile swimmers. This is important, because they are "hunt and kill" predators. They're picky about what they eat, though, preferring to dine exclusively on thecosomes. And not just any thecosome, but their own favorite prey species. The relationship between a given species of gymnosome and the species of the thecosome it eats is so tight that the species name of the gymnosome is often the genus name of the thecosome it prefers to eat. For example, the gymnosome *Clione limacina* (figure 11) specializes on thecosomes of the genus *Limacina*. This specialization even goes beyond names. If you've ever eaten a snail, you know that the trick is getting it out of the shell. Conch fishers in the Caribbean, for example, get the conch out of the shell by creating a small opening near the attachment point of the muscle that holds the animal inside, and then cutting that muscle, causing the conch to fall out of its shell. Animal predators of snails often drill a small hole and eat the snail through the hole, though some—like the mantis shrimp—use their powerful club-like appendages to smash the shell open. Snail shells, even the lighter and thinner ones of pelagic species, are tough and can be hard to get into.

The gymnosomes' answer to this problem is to have a Swiss-army-knife set of tools inside their heads. Although from the outside these animals are cute little slugs with wings, and often have evocative names like "sea angels," inside their head is a set of contraptions that are adapted to get their favorite snail prey out of its shell. *Clione limacina*, for example, first grabs the *Limacina* snail with a pair of soft tentacles. It uses these tentacles to maneuver the snail into just the right position. Then, long, sharp hooks are squeezed out of sacs on its head (this is known as "eversion"). These hooks—much like cocktail forks—can reach far enough into the shell of the *Limacina* to pull the soft body out. From the *Limacina*'s point

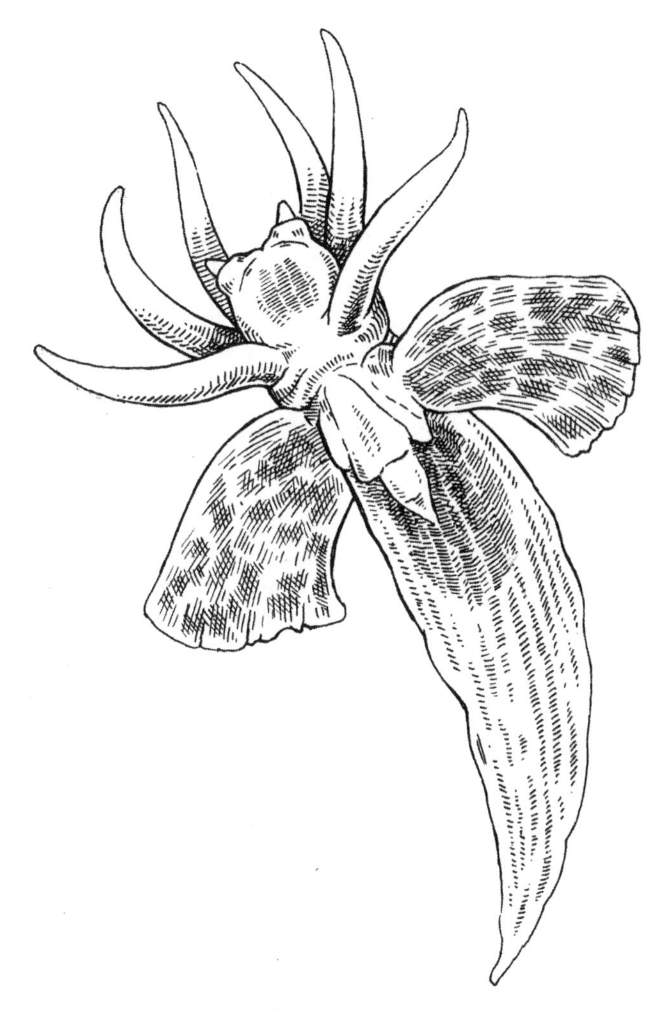

FIGURE 11. The naked (shell-less) pteropod *Clione limacina* with its feeding tentacles out

of view, it must be a nasty way to go, being yanked out of your cozy home and devoured.

The problem with specialist predation of this sort is that the predator is dependent on the prey. In the case of *Clione*, this can be a problem because the abundance of *Limacina* varies strongly with season. These polar animals are abundant during the summer months; other times of year they're rare. *Clione* deals with this in two ways. First, it can eat other animals, and we know that it does so from looking inside its stomach. It can also last a long time without any food at all, slowly digesting itself for energy. Nearly all animals can do this, but since bones don't shrink, vertebrates like us just get skinnier. Invertebrates, which don't have bones, usually get smaller. This is the classic trade-off of a specialized diet. It can allow for the evolution of highly effective eating strategies, but you become dependent on your limited diet, much like my daughter Zoë, who lived off hot dogs for a couple of years in grade school. Much of the United States experienced dependency in 2022 owing to the baby formula shortage. Human infants, along with most other mammals, are milk specialists and are very effective at deriving energy from it. But this leaves them vulnerable to milk or formula shortages, a problem for which the only solution is to end the shortage.

## Not Becoming Food

As hard as it can be to find and capture food in the open ocean, it is just as hard to avoid becoming food. In an environment where food is scarce, where there is nowhere to hide, and where the predators are probably faster than you are, most animals must find new ways to protect themselves. One can, of course, evolve to become larger, and it's possible to imagine a dinosaur-like evolutionary arms race where oceanic animals get bigger and faster. But this is not common. Yes, there are whales, sharks, and some large and powerful fish and squid, but the majority of species in the open ocean do not appear to be evolving to increase in size.

Some animals protect themselves by being toxic or venomous, or simply tasting awful. We've already discussed the various

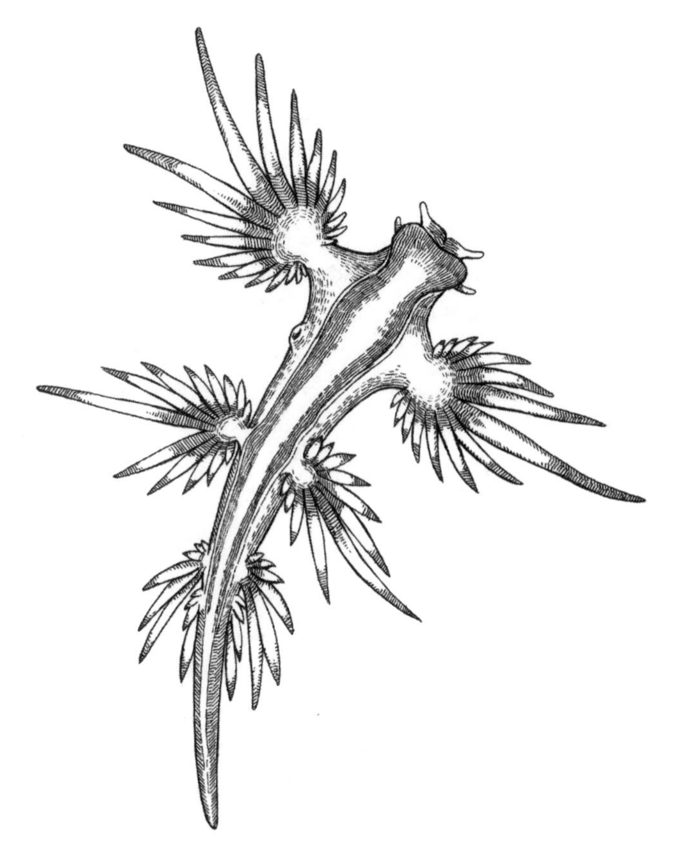

FIGURE 12. The blue sea dragon *Glaucus atlanticus*

jellies that use nematocysts, but it turns out that certain animals can also use nematocysts by stealing them from the jellies. The pelagic sea slug *Glaucus atlanticus* (figure 12) is known for this. It's also known for being one of the most gorgeous animals in the sea. It's only an inch long, but it is a perfect silver and blue dragon, with sets of spike-like appendages that radiate out from pairs of what appear to be arms and legs. I've never shown this animal to anyone who didn't immediately fall in love with it, and one glassblower on the Caribbean island of Saba loved it so much that she made me a glass version, which sits in my office.

*Glaucus* preys on the Portuguese man-of-war, which—as we've discussed—has one of the nastier stings out there. It won't usually kill you as a box jelly might, but for a while you'll wish it had. *Glaucus* not only eats this animal, but manages to ingest the cells containing the nematocysts without setting them off—the equivalent of eating a bowl of hand grenades. It transports these nematocysts to little sacks at the tips of those pretty, spike-like appendages, which concentrates them to even higher levels than are found in the Portuguese man-of-war. *Glaucus* uses these untriggered nematocysts for defense and for attacking its own prey, just as if it had grown them itself. *Glaucus* is not alone in doing this—several other sea slugs use the same trick.

The beauty of *Glaucus* may not be accidental. Many venomous, toxic, and foul-tasting animals and plants are striking. The usual explanation for this is that the organism is warning others of its toxic or unpalatable nature. After all, there's not much point in poisoning the animal that just devoured you, except maybe out of spite. Instead, you want animals to see the colorful patterns and keep away, either because of an innate aversion or because they have learned from past mistakes. For example, I will never eat another lobster sandwich after eating a spoiled one in Nova Scotia. The blue and silver coloration of *Glaucus* may be there to help hide the animal in its silvery under-surface world, but many venomous jellyfish are painted in obvious colors. You might wonder how they catch anything if they're so obvious, but the smaller invertebrates they eat typically don't have sharp enough vision to make out the colorful warning patterns. The larger fish that can see and avoid the jellyfish are the sort of animals that would only damage it if they ran into it.

Despite the value of toxicity, it doesn't seem that many open-ocean animals are protected in this way. Aside from the various medusae and the sea slugs that steal their toxins, evidence from gut content analyses suggests that most animals in the water column of the open ocean taste fine. So, if you can't run, you can't fight, and you're tasty, all that is left to do is hide. We already discussed hiding in the dark in the last chapter when we talked about vertical migration, but what about the animals that don't do this? They need to look like water; they need to hide in plain sight. For the rest of this chapter, we'll discuss how animals pull off this trick.

## Colors

As is true on land and in coral reef habitats, many oceanic animals use skin coloration to hide. There are some differences, though. First, as we discussed in the light chapter, the underwater light field is overwhelmingly blue and varies strongly depending on which direction you look. Second, unlike in coral reefs and other coastal habitats, there is no busy visual background to mimic or disappear into. This affects pelagic animal coloration in a few ways. First, the colors used must account for the fact that they're going to look much bluer underwater. For example, if you want to hide against the blue water, you don't want the color of your skin to match the blue water because the blue light hitting it will make it even bluer, and you will no longer match your surroundings. Second, because light intensity and color vary with viewing direction, your body must have a gradient of color as it goes from top to bottom. An example of this is seen in the bluefin tuna, which is so dark a blue on top that it is nearly black. This fades to a royal blue as you start to move down the body, becoming a sky blue around the midline and fading to silvery white as you approach the belly. Put this gradient of colors in the oceanic light field and the tuna wearing it is well hidden.

A second effect of the blue illumination is that red light is almost nonexistent (see chapter four). This means that red and black look the same once you get below about 60 feet. That might seem unimportant, but many invertebrate animals don't make black pigments. The classic black pigment is melanin, which in the open ocean is mostly made only by fish, sharks, and marine mammals. But because the ocean is so blue, all the other animals can appear just as black using red pigments that they can either make or get from their diet.

Appearing black is more important at night and as depth increases. This is because bioluminescence, which we also discussed, becomes the dominant form of light. This emitted light (which is nearly always blue) is often used to illuminate prey, which means that prey need to reflect as little of it back as possible. Even predators need to worry about this. For example, angler fish, which are famous for the bioluminescent lures that hang in front of their toothy mouths, would not be as successful if the lure also lit up their leering faces. So, a number of species, especially certain groups of fish, have evolved to become ultrablack and reflect almost no light at all. Many fish also have ultrablack guts to hide the bioluminescent food they eat.

Karen Osborn, at the Smithsonian's National Museum of Natural History, and I got interested in these animals after they frustrated our ability to photograph them. She and I typically photograph everything interesting we find at sea, using the images for science, outreach, and conservation, but in every image we took of these fish they just looked like black holes. After many failed attempts, we learned how to measure how much light these animals reflected and were astonished to find that they were as black as the blackest known artificial substance,

Vantablack. If you've never heard of Vantablack, go look up some photos of what objects coated with it look like. In the cases of the fish and Vantablack, only one photon of every two thousand comes back to the eye. This makes them about a hundred times blacker than the blackest objects of your day-to-day life, which explains why we couldn't photograph these fish.

Although coloration is used to hide in the open ocean, a number of other tricks exist in this habitat that are rare or absent in others. These include mirrors, lights, and transparency.

## Smoke and Mirrors

A surprising number of pelagic animals, especially various fish and squid but even some copepods and pelagic snails, use mirrors. Among my memories of the North Carolina beach vacations that changed my life are those of the trips to the harbor at the Oregon Inlet Bridge. The Herbert C. Bonner Bridge itself was impressive, tracing a 3-mile-long arc across water, beach, and marsh, reaching 100 feet tall in places, but it was the harbor next to it that intrigued me. It consisted of mostly private charter boats that took people out to fish in the Gulf Stream for the day. If you arrived at about five p.m., you could see the boats coming in and unloading what they'd caught. In addition to being larger than I thought fish could get, many of the fish looked like they were made of silver.

It wasn't until I was thirty-six years old that I gave much thought to why open-ocean fish were silvery. If it crossed my mind at all, I assumed that schools of silvery fish would be a confusing sight for a predator, making it hard for them to single out one fish to attack. This may be true for silvery fish that are near the surface, where the beams of sunlight glint off their contours, but it turns out that these biological mirrors are primarily used for camouflage. This was first discovered by Eric Denton in the 1960s at the Plymouth Marine Lab on the English Channel, where he was known to walk around the campus holding a dead silvery fish, turning it this way and that against the sun. The first thing to know is that the underwater light field is symmetrical around the vertical axis. Suppose you are a scuba diver 60 feet down. If you were to look up at an angle of, say, 45 degrees above the horizontal and slowly rotate in a complete circle, the brightnesses of the patches of the water you would see during this pirouette would all be about the same. This is true for any angle. For example, it would also be true if you looked horizontally instead of 45 degrees up. Of course, you might say, "Shouldn't it be brighter when you're looking toward the sun?" But as long as you're at least 50 feet below the surface it, isn't—at least not by much. This is for two reasons. First, light bends when it enters the water. You can prove this to yourself by putting a pencil in a glass of water and looking at it from the side. Because of this, if you're underwater looking up, the entire dome of the sky is compressed fourfold into an overhead circle called Snell's window. So even if the sun is low in the sky, the compression still puts it close to directly overhead. Also, as you get deeper, the underwater light becomes more diffuse. So, where the sun is in the sky matters less than what the shortest trip is that light takes from the surface. This shortest trip is always directly down, so as you get deeper, the brightest spot moves to be directly overhead.

The second thing you need to know for mirror camouflage is that light always leaves a mirror at the same angle that it hit it. For example, if light hits a mirror at 45 degrees to its surface, it will leave the mirror at 45 degrees, though in the opposite direction. This is why you can see yourself in a mirror only if it is right in front of you. So, now imagine you are looking at a vertical mirror in the ocean. Let's say you are little below it, so you are looking up at it at a 45 degree angle. Because of the way mirrors reflect light, the patch of ocean you're seeing in the mirror is still 45 degrees up, but instead of being in front of you, it's up behind your head (it's helpful to draw this on a piece of paper). Because the underwater light field is symmetric, though, this patch of ocean up and behind you is just as bright as the patch you would have seen if you could look through the mirror. So, a vertical mirror in the ocean tricks you into thinking that you are looking through it.

The problem is that most fish don't have flat vertical sides. A few do—for example, the lookdown fish, which has a comically disdainful appearance and is flat as a pancake. Most fish, though, are roughly cylindrical. As I tell my students, life is mostly about being a fat cylinder. What Denton discovered, though, was that even though the bodies of silvery fish were curved, the mirrors were still vertical. These small mirrors are made of stacks of thin sheets of guanine, which is an important component of DNA, and are embedded in the scales at different angles, depending on where the scale is on the body. The orientations of the mirrors relative to the scales are just right, so that the whole fish body—even though curved—acts as a vertical mirror. This vertical mirror isn't flat, because it's still attached to the outside of a curved body, so it's best to consider it as a collection of tiny vertical mirrors.

It's not just fish that use mirrors to hide in the ocean. It's also not always the outer surface that an animal is hiding. There are many squid that are transparent, another camouflage trick, which we'll get to soon. This allows them to hide not only skin and muscles but also some of their internal organs. The problem, though, is that they need to eat, and even transparent prey typically become opaque after being chewed up. Even worse, many prey are bioluminescent and will continue to emit light after being swallowed. At night or at depth, swimming around with a glowing stomach is about as dangerous as it gets, so many pelagic animals have opaque stomachs. One way to have an opaque stomach that is still hard to see is, again, to cover it with lots of little vertical mirrors. The squid take this even further by having what is called a "needle-shaped" stomach. It's not truly shaped like a needle, more like a single grain of rice. This makes the stomach harder to see both from the side and from below. Squid go the extra mile and attach muscles to this stomach and use them in a way that ensures the stomach is always vertical even if the animal itself is tilted. Therefore, any animal looking directly from below—which is a common viewpoint in the open ocean—sees the stomach as only a tiny circle. Any animal looking from the side sees (or, more accurately, doesn't see) a vertical mirror. This wonderful set of adaptations is also found in heteropods, a group of predatory pelagic snails.

Finally, mirrors are used to hide eyes. Eyes are a problem because their very function makes them visible. Eyes not only have to absorb light to see, they also need to control where the light comes. If our retinas were glued to our foreheads, we'd be able to sense that light was striking them, but we'd have no idea where it was coming from, except that it was somewhere in front of us. For this reason, most of our eye is opaque, with only a tiny window—the pupil—that lets light in. Eyes in the ocean also tend to be large compared with the size of the animal, so that they can let in more light. And the best camouflage in the world can be undone by the presence of two opaque balls bobbing about. In fact, one of the main ways that we find well-camouflaged animals, both while scuba diving and in a bucket on the ship, is to look for pairs of opaque spheres. Therefore, many pelagic animals cover their eyes with little vertical mirrors that help hide them.

All in all, mirrors are one of the coolest adaptations I've seen, and they're especially well suited to the featureless open ocean. On land, mirrors are rare, being mostly limited to certain scarab beetles, the cause of the eyeshine you see in deer and cats, and the cocoons of a few butterflies. In these cases, it's not even clear if the mirrored surfaces are being used to help the animal hide. But the ocean is full of animals that—while impossibly shiny on the deck of a boat or in a sardine can—are well hidden in their own world.

## Hiding in the Light

Colors and mirrors are great for hiding the top and sides of your body in the ocean. But hiding the bottom is harder. This is because the light going down in the ocean is two hundred times brighter than the light going up. So even if your shiny white belly reflects all the light that hits it, a predator looking at it from below will still see it as black because the light coming down around it is two hundred times brighter than the light bouncing off the belly. In fact, it doesn't matter if your belly is black or white, it will always look black when viewed from below, just like the silhouette of a plane against the sky. Many pelagic predators have evolved to take advantage of this fact, and have eyes that always look up, which is one of the reasons their faces look so creepy.

The solution many fish, crustaceans, and squid have to this "belly problem" is to cover it with a carpet of light organs that together do their best to match the intensity and color of the downwelling light, thus erasing their silhouette when viewed from below. This is known as counterillumination, which we mentioned briefly in chapter 4. You can put a counterilluminating animal in a glass dish and view it from below. If you change the light level above it, the animal will become visible for a second and then disappear as it changes its light emissions to match the new background level. The light organs can be quite sophisticated, possessing lenses, filters, and shutters. In essence, they are like cameras or eyes running in reverse, with the light coming out instead of going in.

The remaining question, though, is, how do the animals know how much light to emit? Their eyes can see the downwelling light, but their light organs are on their bellies, sending light downward. Unlike humans, most of these animals cannot bend to see their bellies, so we don't know how they match the light. We know that many counterilluminating fish have a little light organ near their eye that sends light directly into it. Since animals usually don't want to shine light into their own eyes, it may be that this little light is a calibration system. If its brightness always matches that of the light organs on the belly, the animal—who can now see both the downwelling light and the light from this organ in the same eye—could use it to counterilluminate correctly. We also know that certain counterilluminating shrimp have the same molecules in their light organs that are used for vision in their eyes. So, it's possible that the light organs not only make light but detect it as well. As with many things in pelagic biology, it will take time to figure out how counterillumination works, since almost none of these animals act normally in a lab.

## Life as a Window: Organismal Transparency

The best way to look like water is to be transparent and let the background light pass through you as if you weren't there. It turns out that many pelagic animals are transparent. Of course, we don't know that all of them are doing it to hide. Some could be transparent simply because their bodies are mostly water—for example, the various medusae and comb jellies that are so full of water that there's not much left if they dry up. But there are many transparent pelagic animals that are solid and whose relatives in other habitats are opaque. These include the pelagic snails I love to talk about, along with various crustaceans, squid, octopuses, fish, and worms. They vary in how clear they are, but often the camouflage is so good that we have a hard time finding them on blue-water dives. The typical blue-water collecting dive can look like a frozen television screen, because after we drop to depth we're all just staring blankly outward at the ends of our tethers, a glass bottle in one hand, the lid in the other. On many dives, it can be ten minutes before we see an animal, even though we know they're all around us. One trick we use is to stare at one another. Most of us wear black wet suits, and it turns out that transparent objects are easier to see against a black background. This works great for the smaller transparent animals, which are impossible to see otherwise, but it can be unnerving for a new diver in the group, who is left wondering why we're all staring at them. An annoying situation, and one that has occurred to me many times, is to see the perfect animal right in front of my face, reach down to get a jar, look back up, and the animal is gone. I know it's still there, just a couple of feet away, but I can't find it. The animals can be even harder to see in a bucket on the ship. As I said before, sometimes all we can see are their eyes, even if the animal is many inches long and only a foot or two from our faces. For the animals without eyes, sometimes all we can see are their shadows on the bottom of the bucket as their bodies distort the light that passes through them.

So, transparency is an excellent form of camouflage, but it comes at great cost. Unlike other forms of camouflage, it involves not just the surface of the body but the interior. A pelagic animal using mirrors, colored skin, or counterillumination to hide doesn't have to concern itself with what its organs look like, but a transparent animal does. An even bigger problem is that you can't become transparent simply by removing pigment. Many objects of our common experience, such as snow, milk, and clouds, don't have pigments, but are nevertheless impossible to see through. Similarly, most animals would not be clear if you took away all their pigments. They would just be white, like a person with albinism.

The fundamental problem that transparent animals must solve is reducing what is called light scattering. Light scatters—meaning it bends and reflects—whenever it moves from one material to another. This is a problem because most animals are made of multiple materials. Our bodies are packed not only with organs but also with muscles, blood vessels, nerves, and bones. So, any time light moves between these different tissue types it will bend and reflect, and thus less light will make it through. For example, milk is a mixture of two transparent substances—a clear liquid and millions of little spheres of fat. The fact that light passes between these two substances over and over on its trip through the material is what makes milk opaque (this is also why whole milk—which has more fat spheres—is more opaque than skim milk).

We don't know all the methods that transparent pelagic animals use to stay clear. We know that some do it by being very flat, like the leptocephali we discussed. These eel larvae don't have any red hemoglobin in their blood, which also helps keep them clear. Some animals, like medusae, are thicker, but the bulk of their body is just one nearly homogeneous pile of jelly, known as "mesoglea." Most of the cellular parts of the animal—the parts conducting the business of life—are in a thin layer over the mesoglea, as if the animal were painted on the outside of a clear rubber ball. Certain crustaceans do a similar thing by filling their entire body cavity with a liquid. The best example of this is an amphipod known as *Cystisoma,* whose name literally means, "body like a bag." The living parts of this animal are again mostly in a thin layer on the outside.

Some transparent animals do have living tissue all the way through their bodies, and their transparency remains a mystery. We know from medical studies on the lens and cornea that being transparent is hard. The cells of the human lens get rid of their nuclei and all other organelles to become as clear as possible. The only thing left is a large concentration of a few types of proteins. The cornea is made of the same stuff as the white of the eye: the fibers that are also found in our tendons. The fibers in the cornea are small and stacked like the firewood of an obsessive homeowner. The fibers in the white of the eye are larger and laid down in a sloppy mess. This makes all the difference.

We do know that being transparent makes it harder to be fast and strong. Fast and strong animals are this way because they have a complex inner structure. For example, they have blood vessels that quickly move around food and oxygen, and they have nerves that transmit information quickly. This "indoor plumbing" scatters light, and so transparent animals often do their best to minimize it. Some try to have it both ways. For example, some transparent shrimp are essentially "holding their breath" whenever they rest. By this, I mean that they are not filling their vessels with blood. Shrimp blood is clear, but it still scatters light, so keeping the vessels empty makes the animal clearer. If the shrimp needs to do something, it opens the gates and lets the blood flow to its muscle. We do this as well, which is why our skin often turns red when we exercise, but these shrimp do it to an extreme extent, essentially starving themselves of food and oxygen when they rest. As I said in the beginning, the animals in this pelagic habitat are playing a serious game.

# CHAPTER 7

# Family

> We are like larvae, awaiting the moment when we emerge as our true selves.
>
> —Adam Weishaupt, *The Movement: The Revolution Will Be Televised*

## Whom We Leave Behind

My clearest memory of leaving port on my first research cruise is of my new wife, Lynn, standing on the dock. We'd been married for only three months but had been together for over five years, and she was—and is—the love of my life. Leaning against a telephone pole in white shorts and a teal tank top, she just quietly watched me as the ship moved out of the channel—she was never much of a waver. Just as we turned into the Indian River Lagoon on the way to the open sea, I saw her walk back to our red station wagon and drive off.

I knew I wouldn't see her, hear her voice, or even read a letter from her for two weeks. Ship internet didn't exist in 1997, and the only phone on the ship used a satellite and thus cost too much. The only other way to contact land while at sea was to go to the bridge and use the ship-to-shore radio, which we were told was to be used only for scientific emergencies and deaths among close family. We were similarly isolated from any other events on land. On this cruise, for example, we returned to learn that Princess Diana had died while we were away.

Truthfully, I didn't mind the separation. I was so excited to be at sea that being away from my wife for a couple of weeks didn't even enter my head. At some point during this cruise, I asked Edie how she was doing, and she said, "I miss Dave." All I could think was, "Really? How could you be thinking of your husband when we are diving in a submersible and collecting bioluminescent plankton?" But this was my first cruise, and she had been to sea many times. Later, I understood and joined the many other researchers and crew members wandering the dock looking for a pay phone (or, later, a good cell signal) before the ship left so that they could reach their loved ones one last time. Sometimes I would call Lynn multiple times, even though there was nothing much to say but, "I love you. I'll miss you." Later, when my daughter Zoë was an infant, Lynn would put her up to the phone so I could say something she'd never understand.

These final phone calls could be used to both deliver and get surprises. I once told a university that I would not be accepting their job offer from a pay phone on the dock in Woods Hole immediately before stepping on a ship, so that they couldn't respond. A good friend of mine learned that his wife was pregnant with their first child just before the ship left the dock, from the same pay phone, I think. He spent much of the next two weeks sitting in the lounge of the ship, staring into space and waiting to talk to her again.

Twenty-five years later, the ships now have good internet, and some cruises keep us close enough to land that we can call. A few of us even have satellite phones of our own and can call whenever we like. In some ways, this has made the separation from our families even harder—reading some words or even hearing a voice often only reminds us more that our loved ones aren't with us. We all have things we miss most at sea. For some it's alcohol; for others it's exercise or privacy. For me it's trees . . . and Lynn.

This chapter is about family at sea, which means that it's about reproduction, development, parental care, and the long travels that certain animals undergo to accomplish these tasks. Unlike many of the other topics we have discussed in this book, however, the life histories of pelagic animals are often poorly known. Part of this is due to the relative inaccessibility of the open ocean, but another factor is that life in the water column of the open ocean is by necessity nomadic. There is no reef, field, or swamp to call home. Home is water, and that water is always moving. For this reason, the life cycles and reproductive events of pelagic animals can be scattered over entire ocean basins, making them difficult to sort out. Finally, as we will discuss, many pelagic animals—like the eels from the last chapter—change in drastic ways as they grow, which has made it difficult to piece their lives together.

## The Problems of Sex at Sea

Before we get into the challenges of reproduction at sea, it's important to remember that not all organisms require a partner to reproduce. You may have been taught that bacteria and many other single-celled organisms reproduce by first duplicating their genes (and sometimes other internal structures) and then dividing it all in two. Reproduction performed entirely by asexual means in multicellular pelagic animals is relatively rare. There are some species that reproduce both sexually and asexually, often in alternating generations. But sexual reproduction is typically required at some point in the pelagic realm.

The generally accepted explanation for sexual reproduction is that it increases genetic diversity, which allows organisms to keep up with rapidly evolving pathogens, many of which can create a new generation every twenty minutes or so under the right conditions. Sexual reproduction has several costs, though, possibly the biggest of which is that you have to find a mate. The pelagic environment is especially forbidding in this respect. First, it is enormous in both breadth and depth. Second, the web of currents within it means that nothing stays in place. Third, it is featureless; as they say: "Wherever you go, there you are." Yes, the ocean has gradients in light, temperature, algae, and oxygen, but it does not have destinations.

In this gigantic, empty, and restless world, senses are of limited use for finding a partner, especially if that partner is rare. Vision can do only so much. Even in the clearest and brightest waters, animals with the best eyes can see things only up to about 100 feet away, and these would have to be large things. At depth or at night, bioluminescent signals can be seen from farther away, but it's difficult to identify a possible mate by a flash in the dark, especially since nearly all organisms use the same colors. Imagine a party that is pitch dark and dead silent with only dim flashing lights to tell you that anyone is there at all, and that at this party you must somehow find a life partner.

Smell is also less useful for finding mates than you might expect. The fact that animals in the pelagic world are embedded within currents has a large effect on the utility of smell. If a land animal releases a smell, it can travel on the wind and thus move quickly and often quite far. For example, if the wind comes from the north and I smell a hamburger, I know that walking north will get me to that hamburger. But if I'm being pushed south by the same wind, the hamburger smell won't easily get to me, because I'll be moving too. It eventually will, because smells also move by a process known as diffusion, but that process is incredibly slow. For example, it takes an oxygen molecule almost eight hours to travel a yard through the air by diffusion. In water, it's much slower, and that same oxygen molecule would take about seven years to diffuse a yard. Diffusion is also one of those processes that is nonlinear, which makes things even worse. In this case, diffusing 2 yards instead of 1 doesn't take twice as long, but four times as long (twenty-eight years). Finally, larger molecules (such as the chemicals in hamburger smell or in pheromones that might attract a mate) diffuse *even more* slowly. It's not quite as bad as all this because the animal releasing the smell is probably releasing a tremendous number of molecules. Diffusion is what we call a "random walk," and the seven years to travel 1 yard is only the average speed. Some molecules will travel that yard much faster (and some much slower). So, if the animal is releasing millions of molecules, a tiny fraction of them may get across quickly enough to be useful. We do know that some pheromone receptor systems can respond to a few molecules, maybe even only one. Making this worse, though, diffusion is also random in direction, so even if a pheromone molecule (for example) travels 1 yard, it is almost certainly not traveling in the direction you want it to. The bottom line is that although diffusion is fast and highly efficient over short distances, especially in air, it will not work over distances of many feet. This makes using smell to find mates in the pelagic world difficult for animals that typically are many feet apart.

Sound actually works very well in the pelagic realm. The speed of sound depends strongly on the stiffness of the material it is traveling through, and in water it is about four to five times faster than in air. This speed itself is unimportant, since the speed of sound in air is already tremendously high, but the higher speed in water is closely connected with the fact that sound also travels much farther in water than in air. In addition, the speed of sound depends on the temperature and pressure of the water, and the opposite effects of depth on both these factors create a "sound channel" that has been referred to as SOFAR (sound fixing and ranging transmission). Within this sound channel, which is typically at depths of 1,500–3,000 feet, sound—especially at the lower frequencies—can travel hundreds to even thousands of miles. It is for all these reasons that military submarines do not have windows but instead do all their imaging via highly sophisticated sonar systems. Hearing also is one of the few "directional" senses, which means that most animals who can hear can also tell where the sound came from. This is different from smell, where—unless there is a wind to give you a direction—all you can tell is that you are smelling something, not where it is.

Unfortunately for most pelagic animals, sending and receiving sound require that at least part of their bodies have a density significantly different from that of water, which many of them don't have. As we discussed in the gravity and pressure chapters, many pelagic animals are essentially finely crafted water. The animals that make and hear sound typically have air cavities. Some pelagic fish can interact with sound via their swim bladders, but the true masters of sound in the pelagic world are the whales. The toothed "odontocete" whales (e.g., dolphins, porpoises, orcas, sperm whales) use high-frequency sonar for prey detection and other purposes. The baleen "mysticete" whales (e.g., blue whales, humpbacks, right whales) often use low-frequency sound for communication and sexual interactions. These vocalizations, sometimes referred to as whale songs, can have complex structures that we will likely be puzzling over for decades to come.

Unfortunately, there is little known about hearing in invertebrate pelagic animals. There are a handful of studies showing that pelagic shrimp and jellyfish may respond to or be adversely affected by very loud sounds. Many jellyfish do have statocysts. These are tiny rocks inside a fluid-filled cavity that is lined with hair. Wherever the rocks are sitting in the cavity and bending the hairs tells the jellyfish where down is. Because the rocks have a different density than water, they could be used to sense vibration and possibly hear. In sum, although sound is being exploited in a virtuosic way by a handful of pelagic species, some of which are using it for finding mates, it does not appear to be used in general. There are, of course, many other senses. I teach a sensory biology course, and on the first day have students list senses until they drop. They typically get to about thirty. But most of these (e.g., pressure sensing, gravity sensing) simply aren't useful for finding a mate at distance underwater.

Nevertheless, some pelagic animals do find each other. One of most astonishing animals in the open ocean is the glass octopus *Vitreledonella richardi* (figure 13). Found in tropical and subtropical waters at depths of about 500–2,000 feet, it's about the size and shape of the average octopus you might see on a reef or in an aquarium—except that it is almost perfectly transparent. It even does the vertical stomach trick we discussed in the food chapter; there is video showing it keeping its skinny stomach pointed up and down no matter which way the body is oriented. Even its eyes have evolved in a way that minimizes their silhouette and thus makes the animal even harder to see. The glass octopus is also rare—or, at least, hard to catch in a net. In hundreds of trawls over twenty-five years, I have never seen one. But we do know that they find mates and copulate. We know this for two reasons. First, the male has a special, modified arm known as a "hectocotylus." The modifications—which

FIGURE 13. The pelagic octopus *Vitreledonella richardi*

are found in many octopus species—allow the arm to store sperm. This sperm can be delivered to the female in several ways. In some cases, the arm is inserted into the mantle cavity of the female, where the sperm are released and eventually reach the eggs. In other cases, the sperm are mixed with nutritious substances (known as a "nuptial gift") and handed to the female. And in certain species, the male octopus severs its entire arm via a process known as "autotomy" and hands it to the female, who I assume eats it and gets inseminated in the process. Regardless of the exact exchange, the presence of a hectocotylus in the glass octopus is a good clue that it mates in person. It is of course possible that the hectocotylus in the glass octopus is like our appendix, a vestigial leftover from an evolutionary past when it was used. But we do have another piece of evidence—video footage of two of these living chandeliers mating in the wild. How two rare and invisible animals find each other in such an enormous body of water is one of those things that keeps me up at night. Now and then, an animal does something that seems so impossible that all I'm left with is the nagging concern that we must be overlooking something really big and really important. Lynn and I once stayed in a hotel in McCarthy, Alaska. We arrived in the afternoon. It was a little hazy, but we could see many big mountains around us. The next morning, I woke up to a clearer sky and saw that much of my view was covered by the utterly vast bulk of what I later learned was Mount Blackburn, by far the biggest mountain of the region. I had totally missed it the day before. That's how I feel when I see a puzzle like the mating of the glass octopus—that there's something vast right in front of me that would explain it all, if only the haze would lift.

Some pelagic mating strategies are less miraculous, but nevertheless clever. I had mentioned bioluminescence earlier and dismissed it because nearly all the lights are the same color. But there are other ways to stand out and become recognizable to those who know what to look for. One trick is to use shape. A wonderful example of this is another pelagic octopus, known as *Japetella*. It's small and squat, with a lot more body than legs, and fairly common. Like many pelagic animals, it's bioluminescent, but is unusual in that the bioluminescence is found only in the female and only when she has eggs that are ready to be fertilized. The light organ is a ring around the mouth and is covered with a yellow filter. The blue light coming out of the light organ passes through the filter, making green light. This color might make the bioluminescence more distinctive, but the problem is that so far we've only ever found one cephalopod that can see in color, and it's not this one. But the ring shape is distinctive, and it's thought that male *Japetella* spend their lives looking for what many of us at sea call the "glowing green life saver."

## The Spawners

My dad loves physics. For fifty years, he got up early, walked a couple of miles to the university, did physics until lunch, where he posed and solved physics puzzles with his physics colleagues, and did more physics until dinner, when he walked home likely still thinking about physics behind his quiet eyes in the evening. Because of this, we had a number of monthly science magazines, including *Scientific American,* which in the 1970s and 1980s featured a column by Martin Gardner called Mathematical Games and Recreations. Gardner covered an impressive array of topics, but he was especially fond of discussing Edwin Abbott's 1884 book *Flatland,* which was about life in a two-dimensional realm. Like many books that are exciting for kids (e.g., *Gulliver's Travels*) it was in reality a political satire, but Gardner mostly ignored this and spent his time exploring how life could work in a two-dimensional world. For example, one problem is that a digestive system will cut you clean in half.

I bring this book up because one of the greatest differences between a two-dimensional and a three-dimensional world is that everything in the former is *so* much closer together. So, one way for animals to solve the problem of finding one another in the enormous volume of the open sea is for everyone to go to the surface or the bottom. For much of the ocean, though, the bottom is miles down and subject to extremes of pressure and temperature. The surface is no picnic either, exposing animals to UV radiation, rough seas, predation from birds, and other hazards. The top of the ocean can also be uncomfortably warm, especially when it's been calm and the water hasn't been mixed by waves. So, sending all the animals to the bottom or the surface so that they can find one another more easily may cause more problems than it solves.

Spawning (the releasing of eggs and sperm) is a large and complex field of biology, and there are many ways in which it is done. In many fish and amphibians, the female lays eggs on the bottom, which the male fertilizes. In some cases, one or both of the parents guard the developing embryos; in others the young are left to fend for themselves. Just about every pattern you can imagine exists, but here we are interested in pelagic broadcast spawning, in which a pelagic animal releases eggs and sperm, which are typically positively buoyant owing to the inclusion of lipids and eventually reach the surface. Aside from the 2D/3D advantage, the eggs and sperm are now also in the surface currents, which—as we've discussed—move more quickly than the deep currents. This allows for the eggs and sperm to drift over larger distances, like pollen in a wheat field. A number of bottom-dwelling animals, especially corals, do this as well, so even animals that never move are sending their potential young up and off into the world.

This all sounds very nice, but the mortality is enormous. As we mentioned, and as any shipwreck survivor will attest, the surface is a harsh place. The eggs and sperm are small enough to ignore waves, but they do have to contend with intense UV radiation and predation from just about everyone. The predation is made even more likely by the fact that the eggs are often provisioned with nutrients for the developing embryo, should it get fertilized. Also, the fast surface currents and storms that distribute the eggs and sperm over large distances can wash them up on land or send them to waters too cold or warm for them to survive.

The animals deal with this high mortality in several ways, but the primary solution is one of simple numbers. Broadcast spawning species typically release millions of eggs and sperm each time. A commonly quoted number is that your average oyster releases over a hundred million eggs a year. Many species also typically coordinate their spawning, so that many individuals release their eggs and sperm at the same time. This serves two purposes. First, it makes it more likely that a sperm and an egg from the same species will actually meet. Second, it can overwhelm predation. If the eggs and sperm were released at random over the entire year, predators might always have a meal, but if they're released only on certain days but in profusion, even the hungriest and most abundant predator can't get them all.

We understand coordinated broadcast spawning best for bottom-dwelling animals like corals and starfish. This is an intense field of study with more questions than answers, but it appears that spawning often occurs in the early twilight on one of the days near the full moon. So on certain evenings, the water near coral reefs can be opaque with billions of sperm and eggs from countless species. There are so many that they can't possibly all be eaten, and if they can find a partner quickly enough, they can dodge the dangerous UV rays of the morning sun. The sea is a cauldron of incipient life. The only thing I know of that comes close is what we call "the pollening" in central North Carolina, where for about a month in spring the air is dense with green clouds of pine pollen.

The big question is, among so many sperm and eggs from so many species, all sloshing about randomly, how do the right ones manage to find each other? A common definition of "species" is that two individuals are of the same species if they can reproduce and create a new individual that can also reproduce. There are a few cases where two different species can produce young (horses and donkeys producing mules), but these typically cannot have young of their own. In general, an egg cell and a sperm cell must be from the same species to have any hope of fertilization at all. This gets even trickier when we move away from the coral reefs to the open sea, where animal density drops.

The primary answer to how the right egg and sperm find each other is that the vast majority don't. This is a good thing. Remember that an oyster releases a hundred million eggs a year. If these all got fertilized by a sperm cell, we would quickly be up to our armpits in oysters, even assuming hefty predation on the developing young. The more detailed answer is that the sperm and eggs of each species have special "lock-and-key" chemicals and surface receptors that allow them to recognize each other when they meet. This is another area of active research, most of which is done on corals and other reef animals. What goes on in pelagic species is far less understood.

Not all pelagic animals send their sperm and eggs out unprotected in the hopes that a few will find one another and an even fewer will grow to adulthood. Some do guard their young. This is harder for pelagic animals than it is for bottom-dwelling animals, because protection typically requires a surface. The marine mammals, and a number of other species, get pregnant and thus protect their young inside themselves, but some use other surfaces. We mentioned the animal *Cystisoma* in the last chapter. It has a relative called *Phronima* (figure 14), which looks like a transparent shrimp with four large eyes on its head. This animal's reproductive strategy first involves finding a salp. If you remember from the chapter on locomotion, salps are short, transparent tubes that feed as they move, mostly being

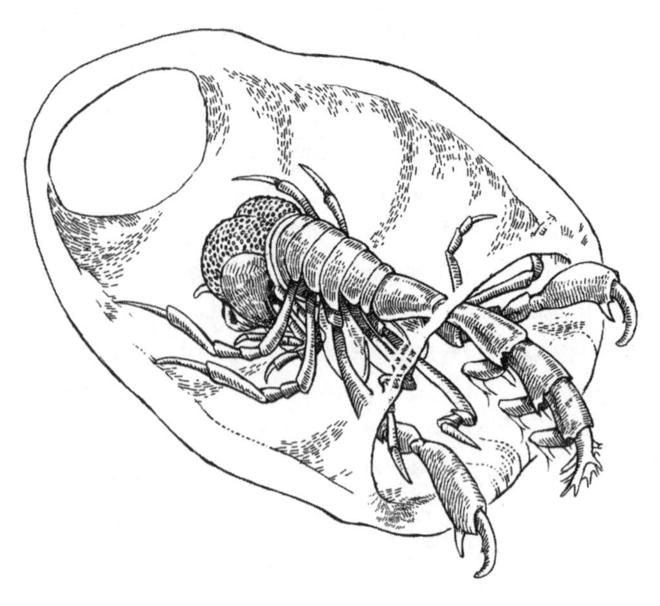

FIGURE 14. A *Phronima* with a salp that it has converted into a mobile nursery

a gelatinous shell with a few organs inside. The *Phronima* takes out all the internal organs (presumably eating them) and, through an unknown process, increases the stiffness of the gelatinous shell into something less like rubber and more like plastic. It then lays its eggs inside the transformed corpse of this salp, pushing the animal around until the eggs hatch. The baby *Phronima* continue to live and grow on the inside of this salp as the adult pushes their home around like a baby buggy. Presumably, they munch on the salp for nutrition.

This parasitic form of reproduction is found in some land animals too. There are wasps that lay eggs inside their caterpillar host, so that the young have something to eat when they hatch. The poor caterpillar is still alive, and in some species the wasp young will eat the less important organs first so that the caterpillar stays alive as long as possible. Even Darwin was disturbed by these wasps, stating: "I cannot persuade myself that a beneficent and omnipotent God would have designedly created the Ichneumonidae [parasitic wasps] with the express intention of their feeding within the living bodies of caterpillars."[1] I think he would have been even more disturbed if one of the parent wasps had come back to push the caterpillar along like a mobile nursery.

## Larvae: The Alien Babies

As I mentioned in the first chapter, I chose to go to graduate school in Chapel Hill because I thought it was close to my childhood beach. What I didn't mention was that I chose to study biology at all only because it began with the letter *b*. I was in the "renovating houses" part of my travels after college, at that time converting an old garage into apartments. The project went very wrong in a number of ways, which eventually led to my friend and I driving rapidly out of Philadelphia in his little hatchback, convinced that we needed to start new chapters in our lives. We didn't understand what graduate school was but decided in a cold minute that this is what we both should do. This left the question of what to study. As we passed the various stores of the local suburban strip mall, nothing came to mind, so we went through the alphabet, where *b* turns out to be an early letter. I had only ever had two biology courses. The first was in ninth grade, where the teacher decided that showing us slides of his vacations out west set to Pink Floyd was more fun than teaching us. The second was a remarkable college biophysics class taught by the equally remarkable Rachel Merz in her

1. Charles Darwin, letter to Asa Gray, May 22, [1860], reproduced in vol. 1 of *The Life and Letters of Charles Darwin*, ed. Sir Francis Darwin (London: John Murray, 1887).

own home, where sometimes her husband would show up and entice their pet turtles to run across the living room floor to attack his wiggling toes. That second class, my love of the beach, and the fact that I had about five tanks of fish in my room in high school led me to believe that biology was the subject for me. My friend stopped at *e* for education, which I always knew he would. Thirty-five years later, we're still living out the decisions we made in that half-hour drive.

I mention all this to say that my commitment to biology was poor at first. In fact, two days before I was supposed to start graduate school, I called my new advisor, Bill, and told him that I wouldn't be coming. A year later, I called him again and said I would. It took me a while to love the subject, but what finally put me over the top—and thus sealed my future—was learning about larvae.

In the normal experience of most humans, we assume that young animals are small versions of adult animals. Puppies look like little dogs, chicks look (more or less) like little chickens, and babies look like Winston Churchill with cuter toes. We do learn that caterpillars turn into butterflies and that tadpoles turn into frogs. Some of us also know that young bees, flies, termites, and ants are pasty grubs, but we usually think of these as the exceptions to the rule that adult animals are just inflated young animals. During the Enlightenment, this was taken to extremes with the theory of preformationism, which believed each human germ cell contained a tiny version of a human, with the "spermists" arguing that the tiny person was in the sperm and the "ovists" arguing for the egg. In case you're wondering, some also believed that each sperm and egg in the tiny people contained an even tinier person.

What I learned in my first year of graduate school was that most marine animals had young that not only looked completely different from the adults (figure 15), but also lived in

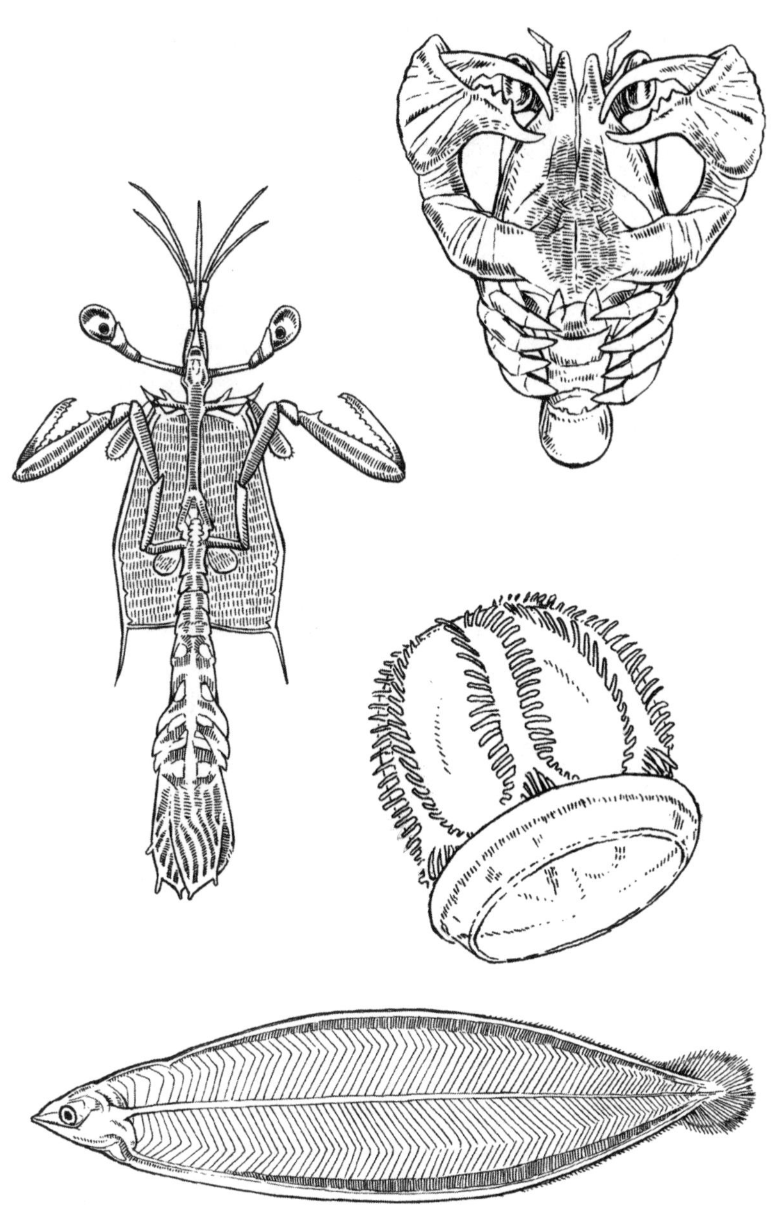

FIGURE 15. Four different kinds of larvae: *left,* stomatopod larva; *top right,* brachyuran crab larva; *middle right,* tornaria larva; *bottom,* leptocephalus eel larva

different places and ate different foods. This is not true of all marine animals. Marine vertebrates typically undergo direct development, which means that the young look more or less like small adults. Very young fish are often called larvae, and some look different from their adult form, but they still all look like fish. Among marine invertebrates, though, larval development is the rule, and direct development the exception.

Three things about marine invertebrate larvae blew my young mind. First, not only did these larvae fail to resemble the adult animal, they also failed to resemble any animal I had ever seen. In fact, they didn't look like any animal I had ever imagined. To my mind, at least, many looked like larger versions of single-celled organisms, like amoebae or paramecia, having no obvious head, blob-like extensions and spikes, and tutu-like fringes made of the cilia we talked about in the locomotion chapter. A few, especially among the crustacea, did look somewhat like tiny shrimp, but with greatly distorted heads and giant eyes. Most were small enough that you needed a microscope to see them clearly, but to my mind they were both grotesque and charming.

Second, although many larvae in the open ocean will turn into adults that continue to live in the open ocean, many are the young of animals that live out their adult lives on the seafloor. The first group are known as "holoplankton," and the second are "meroplankton." The meroplankton in particular got me excited because they undergo "larval settlement," which is when the larva decides that it has found the right place, falls out of the water column, and starts growing on a coral reef, the abyssal plain, or some other seafloor habitat. I had always been a fan of the Weather Girls' song "It's Raining Men," so the first thing I thought of when I learned about larval settlement was, "It's raining babies."

The final wonderful thing about larvae—and this is what turned the key in my head and made me a biologist—is metamorphosis. After settlement in meroplankton (or when the right conditions occur in holoplankton), larvae begin their transformation into adults. At that time in graduate school, I was enamored with brittle stars, which I briefly discussed in chapter 1. They're quicker than starfish, and it's fun to feed them and watch them grab the food in one arm like a football and run off using the other four arms. Their larval form, which is called an "ophiopluteus," always looked to me like a tiny, advanced fighter jet, with swept back spikes and wings. When the larva is ready to transform, it doesn't straighten out the spikes and wings into a star shape while slowly growing a disk. Instead, something very much like a tumor starts growing near the left side of the larva's throat. This "tumor" (known as a "rudiment") has five lobes and is inside a cavity known as a "hydrocoel." The rudiment eventually grows to become the entire animal and absorbs the old larval body for nutrition.

The path between caterpillar and butterfly and between tadpole and frog also has a stage in the middle where nothing seems to make sense. Metamorphosis is the rule rather than the exception in the open sea, though, and there are many variations. One of my favorite life history paths is that of the moon jellyfish. The fertilized egg quickly develops into a small larva, which looks a lot like a paramecium. This eventually settles on the seafloor and transforms into something that looks like a tiny sea anemone. This "anemone" then further transforms into what is called a "strobila." This has a pancake stack of multiple tiny jellyfish called "ephyra," which pop off the top in order in a process known as "strobilation." These ephyra again float around the sea, slowly growing into adult jellyfish, which release eggs and sperm into the water to start the process again.

The trick, of course, for all these larval animals is where to settle and when. As we've discussed, there are few solid surfaces in the water column. Many larvae can metamorphose without a surface, but many require one, especially those that will eventually metamorphose into bottom-dwelling animals. Larvae near the coast will settle at the bottom in the shallow waters. Some larvae will settle on the deep-sea floor. One of the more active areas of ocean research is how larvae find the volcanic vent communities of the deep sea that have been written about in so many deep-sea books. Not only are these volcanic vents far down, but also they don't last long. Although a vent field can last for hundreds of years, any individual vent may stop erupting within a few weeks or months, and the community that depends on the chemicals it emits will die. So, the larvae of these special vent-dwelling animals are under extra pressure to find them. Many ideas have been suggested, but much work remains to be done.

Some larvae specialize in particular surfaces, such as certain barnacle larvae that will settle only on certain species of whales or on floating debris. Many settle on artificial surfaces such as oil rigs and the bottoms of boats. This process, known as "fouling," has led to a well-funded research enterprise in how to prevent it, usually via some form of toxic paint. I go to the beach each summer and live on a marsh, where I keep a kayak. The day I arrive I put the kayak in the water. Two weeks later, when I pull it out, the bottom is covered with barnacles and other little animals.

The *when* to settle question is tightly connected to the *where* in that larvae will settle when they find the right place. What the right place is depends on depth, water temperature, and the physical, chemical, and biological nature of the surface they land on. Once the larva begins the settlement process, it typically doesn't get a second chance because it's glued itself down, so this is again a situation where large numbers come in handy. If a larva lands on a starfish, for example, it is doomed. This is because most starfish are covered with a fearsome collection of tiny Swiss army knives known as "pedicellaria," which kill larvae trying to settle. I suppose the larva must be content with the hope that at least one of its siblings has landed on safer ground.

## Using the Edge; or, You Can Go Home Again

As we discussed earlier, there's a big advantage to dropping dimensions. The broadcast spawners make it easier for eggs and sperm to find each other by sending them all to the two-dimensional surface. An even better solution would be to drop to one dimension and go to the coast, or all the way to zero dimensions and have everyone converge on one spot. Since it's hard to maintain a specific location given the currents of the ocean, the best way to go to one point is to go to land. Not all pelagic animals have this option, because not all can survive on land, upriver, or even at the coast, but the ones that can—known as the natal homers—have a number of advantages.

First, they dramatically increase the chances of finding one another. The classic natal homers are the salmon, who, after a life in the ocean, swim up a specific river to reach the spot where they themselves were born. This concentrates the fish, allowing them to find one another, though it does also give certain predators (e.g., brown bears) the mother of all picnic lunches. Second, and this is less appreciated, by breeding in the place that you yourself were born you're increasing the chances of your own offspring's success. It is the devil you know. If you yourself survived being born on that beach, the odds of your own offspring surviving are higher. Third, land (even upriver) is safer than the open ocean. I mentioned going to the beach each year to kayak. I also go down to the ocean beaches on this island with my daughter Zoë to watch baby sea turtles being born and then make their famous march to the sea. In their trip, which has been dramatized by nature specials, the small but indomitable hatchling turtles brave predation by birds and crabs, dodge pits in the sand, and avoid getting lost in the dunes to walk into waves hundreds of times their height and swim off. Many of the people watching the turtles hatch on our vacation beach must also have watched these shows, because they cheer as the turtles enter the sea and are now supposedly safe. In reality, predation at sea is much worse than on land, and a host of fish and other animals capture the majority of the hatchlings before they even get 100 yards off shore. The beach has given the turtles a relatively safe place to gestate their eggs, however, which is something.

I first learned about natal homing through a chance friendship with Ken and Cathy Lohmann over thirty years ago when I started graduate school. When I got the acceptance call for graduate school I was living in my then-girlfriend's room, covering my part of the rent by fixing things in the apartment. The stipend at UNC was low enough that money was still an issue, especially in the summer, so I was always casting about for extra work. In 1991, Ken needed someone to build a large magnetic coil for him to run experiments. This coil was an open wooden cube about 8 feet on each side, strung with hundreds of feet of red wires going in all directions. It took a summer to build, during which I taught Ken the basics of carpentry and he told me about sea turtle navigation, thus starting a friendship that has lasted decades.

The story for an Atlantic loggerhead sea turtle starts with a mother turtle who, after about thirty years in the ocean, crawls up a southeastern US beach one summer night, much like a tractor, and digs a hole with her rear flippers and—in what appears to be a trance—lays about a hundred ping-pong-ball-like eggs into it. Using the same rear flippers, the turtle refills the hole and crawls back into the sea, leaving tracks that do indeed look as though a large tractor drove into the sea.

Assuming the nest isn't flooded by storms or predated by foxes or raccoons, most of the eggs hatch forty-five to sixty days later, usually between sunset and midnight. The adorable hatchlings enter the sea after the march described above, swim straight offshore, and begin a mysterious part of their lives where they circle the Sargasso Sea over six to ten years as they grow into adulthood. They first roughly follow the Gulf Stream over to Europe, move south toward the waters off the west coast of the Sahara, cross the Atlantic back to the Caribbean, and then move north and east again along the rough path of the Gulf Stream. As they become adults, they move to feeding grounds in coastal waters in the northwestern Atlantic. When they are about twenty to thirty years old they mate in these coastal waters, and the females march up the same beach they crawled down so many years ago to lay their own eggs (give or take some distance up or down the coast).

There are seven species of sea turtles, found throughout the ocean, and so the exact paths of their travels vary, but they all find their way back to their natal beach to lay eggs. Some, like certain populations of the Olive Ridley sea turtle, synchronize their beach arrivals, arriving on a particular beach over a few days in staggering numbers. Others, like certain populations of green sea turtles, find their way back to a tiny volcano in the South Atlantic called Ascension Island.

How do these animals find their way around the ocean for so many years, eventually coming back to the same narrow patch of sand? Ken and Cathy Lohmann have struggled over the last thirty-five years or so to give us a picture of how this works in Atlantic loggerhead and green turtles. Although their initial ability to swim directly offshore appears to depend on their sensing the motion of the waves, it seems that the primary guiding forces for their transoceanic journey is the earth's magnetic field.

Although you may have held a compass in your hands at one point, my guess is that you would not want to be dropped into the center of the North Atlantic Ocean and told to use it to find Baltimore. A compass can tell you where north is (or any other direction), but a compass alone will not tell you where you are. For that you need a map. It turns out that the earth's magnetic field can not only tell you where north is, but also give you some idea of where you are. This is because your typical hand compass doesn't tell you the full story. It turns out that there is a fancier compass built into our cell phones. If you use it, you'll notice two things. The first is that the magnetic field isn't horizontal, but instead points fairly steeply into the ground in the northern hemisphere and equally steeply out of the ground in the southern hemisphere. The magnetic field also doesn't have the same strength everywhere, varying almost twofold over the surface of the earth. Other aspects of the local magnetic field vary too, and they all do so in a predictable manner, such that if you could read them you could tell roughly where in the world you were—much as with city or state names on a map.

A number of experiments in Ken and Cathy Lohmann's lab have suggested that loggerhead and green sea turtles do have this fancier compass. One piece of evidence is that they seem to do the right thing when placed in artificial magnetic fields that mimic those found at critical points in their journey around the Sargasso Sea. For example, there is a critical point off the coast of Europe where, if they keep following the Gulf Stream north, the turtles will eventually reach waters that are cold enough to kill them. We find few dead sea turtles on the western beaches of Ireland and Wales, so somehow the turtles know to make a hard right off the coast of southern Europe to head south toward Africa. From the lab experiments, it seems as though turtles use aspects of the magnetic field to know when to make this turn and other important turns on their multiyear ride around the North Atlantic. Another piece of evidence, more closely related to natal homing, comes from the fact that the earth's magnetic field changes over the years. It doesn't change by much, but over thirty years it's changed enough that the magnetic "signature" of a given natal beach might now be found 10–20 miles down the coast. Recent evidence has shown that sea turtles appear to pay more attention to the magnetic signature of their home beach than its actual location, and end up changing where they crawl up the beach based on these changes in our magnetic field.

The big question, of course, is, how do they know the magnetic signature of their home beach? Ken and Cathy Lohmann have suggested that sea turtles may imprint on their natal beach. "Imprinting" is an old biological term that was first used to describe how certain newly hatched birds might choose a human as their "parent" and follow them around if that human was the first thing they saw. First described in domestic chickens by Sir Thomas More in the 1500s, the concept was expanded and popularized by the animal behaviorist Konrad Lorenz five hundred years later. The concept typically assumed that newly born or very young animals permanently associate certain humans, animals, or objects with important functions (e.g., being a parent) during what is known as a "critical period." It is possible—though far from proven—that sea turtle embryos learn the magnetic features of their home beach while still in the egg under the sand.

On the topic of mysteries, I want to end with a final one on elephant seals. Elephant seals live up to the "elephant" part of their name, with the males reaching lengths of 16–20 feet and weights of 5,000–7,000 pounds—about the length and weight of a large American pickup truck. More remarkable, though, are their migrations and natal homing. The northern elephant seal has a few breeding colonies on the west coast of the United States and Mexico, but one of the most populated is found in Año Nuevo State Park just north of Santa Cruz, California. In December of each year, the males arrive. They are much larger than the females and use their slightly earlier arrival time to fight one another for dominance. The females—which are still 10–12 feet long and up to a ton in weight—arrive a few weeks later, give birth, nurse their pup(s), and mate with the dominant males about a month later. During this entire period of birthing, nursing, and mating, the female seals do not eat, but after mating they go back to sea and head thousands of miles in directions that range from west to northwest. After feeding for a time in what appears to be a random patch of the North Pacific Ocean, they swim more or less directly back to the park for another short time on land, where they undergo what is called a "catastrophic molt" of skin and fur. After this, they go back out for an even farther and longer journey. They not only find their way back to Año Nuevo in December, following roughly the same path that they took out, but they get there just in time for their pup to be born. A pup cannot survive being born at sea, so this timing is critical. The longest known journey of this sort is 13,000 miles round trip, and how the seals get back to the park at just the right time is anybody's guess. They mostly swim at depth of a couple of hundred feet, so they can't see the sky to navigate by the sun or moon. If you've ever tried to swim in a straight line underwater without any guideposts, you'll be lucky to make it about 50 feet. So, the best guess is that the animals are using features of the earth's magnetic field—but these are *not* the kinds of animals that you can put in a tank and do experiments on. You can't even walk near one without putting yourself in danger. This means the solution to this puzzle is going to have to be indirect and clever.

As with most animals, the life of an elephant seal is, "eat, mate, repeat" on a yearly cycle. As with certain animals (e.g., many migrating birds), the eating and mating locations are thousands of miles apart. What makes elephant seals odd, though, is that each seal goes to a different spot to eat. It's as if the Pacific Ocean is a vast fridge and each seal is going for its special shelf of goodies. I recently gave a talk at a local high school and—as so often happens—the students asked me what my favorite animal was. I gave the answer I always do—the one I'm working on now. Each animal is its own piece of magnificence and its own puzzle. Right now, we're trying to figure out how elephant seals navigate. After two years we have gotten nowhere.

# CHAPTER 8

# Community

> In all our searching, the only thing we've found that makes the emptiness bearable is each other.
>
> —Carl Sagan, *Contact*

## The Human Community at Sea

The scientists who go to sea year after year are adept at many things. In addition to being well-versed in their scientific subfield, they're good enough naturalists to know what most animals out there are. They're comfortable working with large machinery at sea, including launching and recovering 100-foot-long nets off the stern while waves wash over them. Many can build or fix just about anything out of cannibalized parts, duct tape, and zip ties, and most have developed an excellent sense of balance and relative immunity to seasickness. Some might call them hybrids of university professors and commercial fishermen, but whenever I look at my seagoing friends, the first though that comes to my mind is—"What a bunch of nuts."

While I'm sure that some seagoing scientists are serious, the ones I've worked with over the last twenty-five years all have a bombproof sense of humor. In some, like myself, the humor is visible and continual to the point where they act like oversized eight-year-olds. In others the humor is hidden, but they're always ready with a hair-curling story. Many of the funniest moments of my life have been at sea.

I don't think this is accidental. Research cruises are high-pressure environments. Due to the enormous cost of the ship, the scientists want to get the most out of their time, so they tend to work day and night with the continual worry that something unfixable will break and prevent them from completing their research. Ships are loud, industrial, always shaking and rolling, and parts of them smell horrid. There is also an enforced intimacy that comes from lack of privacy. People will see those thick glasses that you wear only after midnight, they'll hear you snore in your tiny four-person cabin, and they'll know how you smell after a few days of stormy weather keep you from showering. From the moment you step on the ship to the moment you run away from everyone else at the airport long before the flight home, someone will know almost everything you do. The people on board also come from many backgrounds and hold many beliefs. One division is between the science crew (the people that got the money to do the research) and the ship's crew (the people that work year-round to run the ship). These two groups may come from different backgrounds, but the bigger issue is that while the ship is a two-week hotel stay for the scientists, it is typically home for the crew, at least for much of each year.

Put all this together and the need to bond is clear. Much critical bonding can occur as early as the first day, when the scientists are loading their gear. As I mentioned, oceanographic gear—because it has to withstand pressure—is typically packed in heavy metal cylinders. The precruise loading day always seems to be hot, and the tide always seems to be such that you have to drag these heavy objects up or down steep, slippery ramps, and through sea doors on the ship that are 2 feet off the ground and hard to drag things through. *No one* wants to load the gear, but if you don't, you've lost your chance to bond over sharing the hard work.

Bonding over work continues throughout the cruise, and can include tasks such as pulling up a 100-pound camera by hand for hours each evening because it was lowered down a mile, but I think that humor is the glue that keeps ships happy and functional. Part of this is the "live and let live" self-humor that allows people to meet each other halfway, which is especially important in four-hour dives in tiny submersibles. The rest of it is keeping one another entertained with stories and jokes. Most of this is inside humor of the "you had to be there" variety, so there's no point in repeating any of it here, but it keeps people bonded and helps them forget that they're crammed together with many strangers, far from their loved ones, in a metal box that is the only thing that's keeping them alive.

This underlying reality is of course still there, so in addition to humor, the ability to remain calm in a crisis and the capacity to put the team above all else are also critical. At some point in the last twenty years, someone (maybe in my lab) came up with the lifeboat test. Our biggest fear at sea is fire. It takes a while for a ship to sink, but a large fire can make a ship unlivable in a matter of minutes. I've been on one cruise where an engine room fire was suspected and remember the chief engineer going down each flight of stairs in single leaps. The lifeboat test is: "If the ship goes down in flames hundreds of miles from shore, would you want this person in your lifeboat?"

There's not much to do on a lifeboat but turn on the satellite beacon and pray you get rescued, so the physical skills required for the lifeboat test are minor. But the psychological ones are major and affect the survival of everyone in the lifeboat. Science is typically a safe profession; in fact, some of the longest-lived people in the United States are astronomers and physicists. Oceanography is also—on average—safe, but there is the potential for something serious to occur. I have never been on a cruise where someone was badly hurt, but a number of my friends have, and a few of my friends have been injured. So while the risk is small, it is real and not to be ignored.

So, whenever I meet a person who says they want to work in my lab, I consider whether this person would pass the lifeboat test. Are they resourceful, calm under pressure, and willing to make a sacrifice for the group? Will their actions make a serious situation better or worse? This is, of course, difficult to find out while walking around a safe, leafy campus, and the people that do end up in my lab often joke about the odd questions I ask during interviews.

Leadership is, naturally, also important. About twenty years ago, I was reading every possible book on polar exploration, and the effect of leadership on expedition success and survival was astonishing. One the one hand, you had Ernest Shackleton keeping his whole crew alive for years in the Antarctic after his ship was crushed by ice. On the other hand, I remember a leader exploring the coasts of Greenland who killed one of his crew for eating a shoelace and would hog the heat in their little shack by wrapping himself around the one stove. Research cruises have two leaders. The leader of the science crew, known as the chief scientist, makes all the decisions about where the ship goes and what science will be done when it gets there. The chief scientist also handles disputes between scientists over space, animals, and other issues. The leader of the ship's crew (the captain) is the final word on everything that happens on the ship. In addition to making sure that the ship's crew does its job, the captain must approve of all boat movements, all submersible/ROV deployments, and in fact anything that goes over the side, even a small bucket. The captain has complete control over the ship and can in theory lock someone up in their quarters or force them to leave the ship. I have never seen the former happen, but I have been on a ship that suddenly went to shore to force a crew member off. These are rare instances, though, and for the most part the captain works closely with the chief scientist to make sure that as much science gets done and in as safe a way as possible. In a way, the captain and chief scientist are wedding planners, just with a much higher cost per plate. I have sailed with many captains, and they all pass the lifeboat test, being low-key and (mostly) humorous on the outside, while continually navigating and mitigating potential problems behind their eyes.

From here, following the pattern of this book, I should go into all that we know about animal communities at sea, using what we know about human communities at sea as a rough guide. But I can't, because we don't know much.

## We Don't Know

I was at a work meeting about a year ago and a chemist next to me, in jeans and a sport jacket, told me that "scientists are driven by curiosity and ego." I'd have to agree. The curiosity is related to the obsession I discussed in the pressure chapter. The ego can be about money, prestige, power, but one that is specific to scientists is the ego of always having the answer. The academic scientists among us are, after all, called professors, so we feel that we can and should go about professing—that is, telling people what we know. There's a sculpture of three scientists sitting on a bench in between two major scientific institutions in Woods Hole, Massachusetts. The joke is that all three are talking, but none is listening. Authors, including me, are perhaps the worst, cajoling people with hopefully entertaining prose to read about all the things we know (or at least that we could find out in time to write the book).

I'm generalizing horribly here, but a field based on training people to accumulate and disseminate knowledge has likely created many people who want to tell other people what they know. While good in the classroom, this trait can be hard to turn off, which can make us domineering in conversation and difficult spouses. As a friend of mine said to her husband: "Can you just for once stop being a scientist?"

A larger issue is that it can be hard for scientists to admit what they don't know. We can be like politicians, answering a question we do know the answer to rather than the one that was asked. It can be hard to get a scientist to say, "I don't know," and even harder to get them to say, "I was wrong." This is a problem because the scope of what remains to be known is beyond comprehension. If a pea is what we know, the plate it is on is what we thought we knew but got wrong. The ship containing the plate is what we don't know, and the sea around the ship contains all the things we don't even know to ask yet.

This is *not* to say that scientists are wrong about major issues. Important topics (e.g., evolution, the existence of climate change, the dangers of high blood pressure) have been studied by tens of thousands of scientists for decades—we are certain about them. But when you hear a story about two unusual open-ocean animals interacting, assume that we don't have the whole story yet, because they may have been seen only once by only two people.

My advisor Bill had a poster in his office where the wing patterns of various butterflies spelled out Theodore Roethke's quote, "All finite things reveal infinitude." I loved the poster and the phrase, and do believe that we can learn a lot about the infinite from studying what little we can. The infinite still exists, however, and even a short walk in the woods behind my farm confronts me with the reality of all that I'll never understand. Almost daily, I realize that I have no idea what's truly going on. I suppose I'm in good company. Isaac Newton, in the latter part of his life, said:

> I do not know what I may appear to the world, but to myself I seem to have been only like a boy playing on the seashore, and diverting myself in now and then finding a smoother pebble or a prettier shell than ordinary, whilst the great ocean of truth lay all undiscovered before me.[1]

This is a long-winded way of saying that what we don't know is just as important as what we do know. And we don't know much about pelagic communities. So, this chapter, in honor of all that we don't know and will never know, focuses as much on questions as on answers.

## Why We Don't Know It

The problem with understanding pelagic communities is that it involves studying the behavior of animals without disturbing their behavior. We briefly touched on this issue in the light chapter when discussing why we can't figure out what many of the uses of bioluminescence are, but we didn't get into what a

1. Quoted in Joseph Spence, *Anecdotes, Observations, and Characters, of Books and Men* (London: John Murray, 1820), 159. It is reported that Newton spoke these words shortly before his death to Sir Andrew Michael Ramsay, but Ramsay is recorded as being in France at that time. Whether he truly said these words is probably lost to time.

profound issue this is. As I mentioned in that chapter, we can investigate the social life of terrestrial or coral-reef animals by setting up remote cameras or simply by hiding in the brush. Much of what we know about the behavior of birds and primates (just to name two groups) comes from these techniques. There can be problems getting the cameras and people into place, but these have typically been solvable. In the open sea, we're not sure these issues are solvable, even in principle. This is because many of the traits of the open sea that we've been discussing in this book—its harsh physical conditions, its enormous size and emptiness, and the impossibility of hiding—make observations of undisturbed behavior difficult to obtain.

For generations, what we've known about the water column of the open sea has come from dragging nets through it, first via commercial fishing, and later via scientific sampling. This is good for determining what is in the water, though, as we said, it mostly catches the animals that are unable to get out of the way. It does a miserable job of determining the behavior of the animals before they were caught in the net.

Suppose an enormous bulldozer plowed through an open-air concert, scooped everyone up, and dragged them 10 miles to a group of anthropologists and dumped them out. The scientists would be able to tell who had been at the concert, their heights and weights, maybe their ages and a few other things. But they would not be able to say who was sitting next to whom, let alone how they were acting right before they were swept up. They wouldn't know who was there with a date, who was in the process of buying hotdogs for their children, or who was plugging their ears because the music was too loud. It's a disturbing metaphor, but it's worth realizing how disruptive net sampling is, and how little about animal behavior you can learn from it. Rich Harbison at Woods Hole Oceanographic Institution said net sampling was like dragging a grappling hook over London and expecting to understand English society.

Oceanographers of course realize this; which is why they've developed other sampling machines and methods, including submersibles, ROVs, and blue-water scuba diving. Submersibles are wonderful tools, in that they let us go into the habitat and observe animals that haven't been damaged by net sampling. But one cannot say that the animals are undisturbed. Submersibles come in many forms and sizes and serve many functions, but the ones that I have used are about the size of a small school bus, have about ten large propellers, and periodically belch air from their ballast tanks so that they can change depth. Worst of all, at night and at depth, they use bright lights. Certain animals found at deeper depths are permanently blinded by these lights, and nearly all are either startled or driven away by them. A few are strongly attracted to the lights and buzz around the submersible in swarms. Some, like swordfish, will attack the submersible. Many animals, however, just freeze in place, likely feeling something akin to stark terror at the large, noisy, and bright interloper in their world.

Also, submersibles can stay down only so long. This isn't so much due to oxygen limitations as it is to battery life and passenger comfort. The battery of the *Johnson-Sea-Link*, for example, is about as long the submersible itself and weighs over 3,000 pounds. This gives the submersible about four hours of power, less if it needs to use the propellers a lot. I've seen the submersible come up at night without power, and it's a scary event.

ROVs don't have humans inside and get power via a long cable that is attached to the ship, so they can stay down as long as the pilots in the control room on the ship are happy to drive them around. They are still major underwater presences, though, and so are likely to change the behavior of the animals, especially those that can detect light and/or sound.

In the 1970s, blue-water scuba diving was developed to get around some of these issues and allow scientists to study undisturbed behavior. It's a remarkable technique, which I've discussed many times in this book, but divers in their wet suits, continually exhaling bubbles, are still an intrusion into what is otherwise a quiet and empty habitat. Also, because any sort of dive accident at sea leaves the divers far from the rapid help they need, we tend not to go deeper than 80 feet.

So, scientists have developed stealth camera systems that are deployed either on the bottom of the sea or in the middle of the water column. The ones at the bottom are held down by gravity; the ones in the water column are usually hanging from floats via miles-long ropes. Some are towed behind ships. Many are designed to be as undetectable as possible, drifting black and silent for days far beneath the waves. The Medusa camera system, which I discuss briefly in the epilogue is designed to be as stealthy as possible.

The problem with these systems is that they're stuck in one place at the bottom or drift at random. Because the density of animals in most of the ocean is extremely variable, and on average quite low, you can recover your camera and find nothing on it. On one cruise, we used a stealth camera system that sat on the floor for days at a time, saving only short recordings when something moved in the field of view. When we brought it up, it was my job, along with Erika Raymond, to look at the hundreds of thirty-second snippets to see if the camera had recorded anything interesting. But 95 percent of the clips had recorded only because a string with a knot called a "monkey fist" floated back and forth in the field of view. Erika and I would compete to see who could yell "monkey fist!" before the other.

The solution to the low animal density is to lure the animals in somehow. Our early solution was to zip-tie dead and rotting fish to the camera system before each deployment. Later, Edie developed a false bioluminescent lure using blue LEDs arranged in a circle. It looked just like a restaurant pager, but did an excellent job of occasionally bringing in a big animal. The problem with this, of course, is that we still weren't recording undisturbed behavior. Instead, we were recording what animals do when they find a pile of food, which is . . . eat it.

How little we know about the sea was truly driven home to me on a cruise about fifteen years ago. I was in a small dive boat about a quarter-mile from the main ship with some other shipmates and my new student, who had never dived in the open ocean before. We all jumped in but couldn't descend because my student was still organizing his equipment. I then noticed that the bow of the main ship was full of the rest of our crew, all looking our way. Some even had binoculars. Then the radio came on and the person driving our boat heard that there was a 20-foot-long dark creature circling our dive boat. So we got out of the water. Given where we were, this was likely a basking shark, which was both harmless and exciting. Or it could be a large predatory shark. So, going back in could have meant either the dive of our lives or the last dive of our lives. I was tempted, but thought of my wife and daughter and called off the dive. Despite the animal being only a few feet away and huge, we never did find out what it was.

## Associations between Different Species; or, "The Thing That Wouldn't Leave"

My wife Lynn is an immigration lawyer. Over the last thirty years we've been together, I've learned two things. The first is that the law doesn't make sense, which eventually saved me from many arguments. The second is that lawyers and scientists both have many special definitions for words with otherwise common meanings. In biology, we have, for example, "fitness" and "adapted," which seem like normal enough words until you lob them into a discussion with evolutionary biologists and the hair splitting begins. The relevant word here is "community." When biologists talk about "community" and "community structure," they are not typically talking about social behavior. Instead they are talking about the different species in a habitat, which ones eat which, which ones are more successful at populating the habitat, and how this might affect resources flowing into and out of the system. This first section on associations is a bit more like the biological definition of community, in that it explores who is connected to whom.

A great advance that was made possible by submersibles, ROVs, and blue-water scuba diving is that we can now see who lives closely together, rather than trying to guess from the messy assemblage of animals in a trawl bucket. The animals we see may still be quite perturbed by our presence, but if two of them are seen together we can reasonably bet they're together at other times. And the connections we see are remarkable in their diversity. We see octopods holding on to salp colonies and jellyfish, anemone-like creatures called hydroids living on pteropod shells, smaller fish (including remoras) surrounding larger fish and sharks, tiny fish surrounding larger jellyfish, and even anemones on jellyfish. The majority of the associations that have been discovered are those between crustaceans and gelatinous animals such as salps, siphonophores, and jellyfish. Certain crustaceans, such as the hyperiid amphipods, which include the *Cystisoma* and *Phronima* we discussed, and the isopods, seem to specialize on gelatinous animals. Other groups of crustaceans, such as copepods and barnacles, have members that specialize in living on fish and whales. So, while there are a variety of associations, they do seem to follow a few pairings of major groups. Also, it appears that one of the two animals in the association is usually much smaller than the other and is apparently "hitchhiking" on the other. An exception is the hyperiid *Phronima* with its salp baby buggy, both of which are about the same size. Another exception, also a hyperiid, is *Hyperiella,* which carries around a toxic pteropod so that it itself will not be eaten, a strategy not so different from a person carrying around a bowl of rotting meat to clear the way in a crowd.

The first question in any association is, "What is the benefit?" The first answer that comes to my mind is related to the dearth of surfaces in the pelagic ocean. Put literally anything into the ocean—for example, a paper cup—and within thirty minutes, small fish and crustaceans will start swimming around it. Within a few days, another animal may lay eggs on it, and within a few more days, various larvae will settle on it, metamorphose, and begin the surface-bound portions of their lives. Surfaces are scarce in the open sea and many of the few that exist are defended by chemical and other means. So it appears that many of the smaller, hitchhiker animals are using the larger animal as a surface. This surface comes with many benefits. Buoyancy is less important and transportation is free, both being handled by the larger animal. Some of the hitchhikers may also be protected chemically by the stinging cells of the jellyfish or siphonophore that they are now living on. At the very least, they may now be better hidden from predators. They may also now have a place to lay eggs and raise their young.

In short, the hitchhiker gets a good deal in the association. The question is, what does the larger host get? There are three forms of associations: symbiotic, commensal, and parasitic. In the first, both animals benefit; in the second, one animal benefits and the other doesn't care; in the last, one animal benefits and the other animal is harmed. There is a long history of dividing associations into these categories.

Dakota (Cody) McCoy and Eleanor Caves are associated with my lab, the first a current member and the second a member in the recent past. They're both creative and impressive researchers, but have very different views on associations. Eleanor, who studies the cleaning relationship between certain reef shrimp and their fish "clients," is a strong believer in the existence of behaviors that are of mutual benefit. Cody, who studies both reef corals and human pregnancy, believes that there is an evolutionary struggle between all associated animals, no matter how cooperative they appear on the surface.

When it comes to pelagic associations, I side with Cody and think that—at least in the case of gelatinous hosts—the host does not benefit and is typically harmed. At a minimum, the host is now heavier and thus needs to work harder to move forward and not sink. The garbage-can-sized jellyfish *Deepstaria* might ignore the weight of its isopod associate, but most jellyfish likely can't. The anemones that live on jellyfish and the copepods that live on fish are directly feeding on the poor hosts and are thus parasitic. In the case of smaller fish surrounding larger fish and sharks, the cost to the host may be minor, but I'm unsure if there is a benefit. Remoras, which attach themselves to larger fish and sharks and thus get a free ride to a meal that is already killed for them, are thought to benefit their hosts by eating parasites off their skin, much in the way of cleaner fish and cleaner shrimp. I'm unsure, however, if this counterbalances the extra effort the host has to exert to swim with a fish stuck to it.

Much remains to be discovered in this field, but my guess is that many of these associations will be found to range from minimally parasitic to strongly parasitic. It's perhaps not surprising that the more mobile of the two animals, and thus the one that can choose to end the partnership, is typically the one getting the net benefit.

## Social Behavior within Species

There are of course things we know about the social behavior of pelagic organisms, particularly in the larger vertebrates, such as large fish and marine mammals. Blue marlin, certain other pelagic fish, and even some squid are known to pair-bond, meaning that male-female pairs stay in close association for long periods and possibly mate multiple times. Cooperative hunting has been seen in tuna and billfish (the group that includes marlins and sailfish). These fish are even known to rapidly change their skin color, presumably signaling to each other as they herd their schools of prey. Some pelagic species migrate in groups, including the basking sharks, which can travel entire ocean basins in close association with one another. Finally, a fair bit is known about the social behavior of whales and dolphins, most of which are pelagic. Since there are excellent books on this topic, I won't cover it here. We also know a fair bit about the social behavior of seals, sea lions, and walruses, but mostly from their activity on land, which appears to be where most of their social activity occurs.

When it comes to the vast majority of pelagic animals, though, the most we know about their social behavior is whether they come in large groups or not. Some lonesome species are nearly always found by themselves, and others can be found in abundance but without any obvious groupings. Some are found in schools, which can range from just five squid darting about together behind our boat at night to tens of thousands of fish in a tight clump that appears to move like one giant amoeba. There can also be copepods and shrimp in tremendous abundance over large swathes of the ocean, and indeed there is a gray area between a very large school and an abundance of animals. For me, the dividing line is that a school has to move together as a unit.

One evolutionary explanation for schooling is based on what is known as the "selfish herd" theory. In this theory, being part of a large school means not only that a predator is less likely to choose you for its lunch, but also that you can put your fellow schoolers between you and it. This is similar to the common joke that you don't have to run faster than the bear that is chasing you, just faster than your hiking partner. Unlike predation on land, however, in the pelagic world smaller prey animals cannot typically outrun their predators, but they can at least hope that someone else will get eaten first. The downside to this strategy is that a large school is easier to see than the individual fish, and thus will attract more attention. We studied this a few years ago and found that—because even clear oceanic water is much murkier than air—even a huge school isn't visible underwater more than a couple of hundred feet away. However, a large school is still highly visible from the air, so fish that school near the surface will have to deal with predation from birds. Also, as we discussed in the food chapter, in some cases the entire school is eaten, at which point the selfish herd theory won't help you. Schooling has many other advantages, though, including more efficient swimming and better ability to sense predators. And because stragglers that fall out of the school are more vulnerable to predation, it can also increase the average health of the fish population.

Given how organized a fish school looks and how quickly it responds to any startling event, it's tempting to assume that the groups have leaders within them that every fish pays attention to. This has been studied by mathematical biologists, including my current postdoc Jesse Granger, who have found that you can get organized school behavior even if the members all pay attention to only the few animals right next them. This is a form of what we call "collective behavior," and is an exciting area of research.

The schools, pair bonding, cooperative hunting, and group migration tell us that pelagic animals must be communicating with one another. This is not surprising. After all, our world on land is rife with communication. The calls of birds, the patterns on butterflies, and the rhythmic courtship dances of certain spiders all communicate information. Although few pelagic animals appear to make sound, and visual communication is more limited because the water is darker and murkier than air, the behavior of certain pelagic animals suggests that they are communicating with one another, likely via color and lights. Earlier, I discussed the Humboldt squid, which flash their entire bodies rapidly back and forth between red and white. These animals have light organs behind their colored skins that allow these flashes to be obvious even in dim illumination. The question is, what are they saying to one another? I joked earlier that they're just yelling, "My fish! My fish!" but the reality may be subtler and part of a cooperative hunting strategy. We do know that octopods, which are relatives of these squid, are about as smart as cats, so cooperative hunting via color signals is not that big a stretch.

This is the puzzle. We know that squid, pelagic octopods, and many pelagic fish are remarkable at changing color. We also know that all of these animals, plus many of the crustaceans and a host of other animals, can emit light of their own. We've also known for many decades that the patterns of light organs on the bodies of these animals often depend on exactly which species they are. The fish and squid, in particular, seem to have species-specific maps of lights on their bodies, each looking like a different town seen from an airplane at night. We also know that coral reef squid, fish, and crustaceans use color to communicate with one another, as in the ostracods' bioluminescent courtship display that we discussed in the light chapter. Oceanographers have known all this for decades. However, owing to the challenges of getting undisturbed behavior from pelagic animals, we can't be certain that they are actually communicating with one another.

I've always loved archaeology and ancient languages, and was especially fascinated by the Phaistos Disc, which was found in Crete in 1908. It's a clay circle, about 6 inches across, and has a spiral of about 120 complex symbols on each side. Nobody has been able to decipher it or even figure out what it is. Each pelagic squid or fish, with its special and intricate pattern of light organs, is like that disc to me. I know something interesting is there, but I don't know how to find out what it is.

Our knowledge of the social behavior of pelagic animals may have to wait until we have observational methods that are even more stealthy. This is only half the problem, though, because our camera has to be able to follow the animals around the ocean for long enough for something to happen. Given that everything in the ocean is moving all the time, and many animals are rare and possibly long-lived, this seems almost impossible to me.

# Epilogue

One of the things I say over and over in talks, in classes, and in the lab is, "If you want to understand an animal, see what it is like in its habitat." So many times people observe a trait in an animal in the lab and then try to explain why it matters to that animal without considering where it lives. One example are the iridescent comb rows found on comb jellies, which we discussed in the motion chapter 8. They're stunning when photographed in a lab with a flash, so you might assume they have some visual meaning. But when you dive with comb jellies, the ambient light is different from that in the lab, being much more diffuse, so the comb rows seldom look iridescent. For this reason, I often follow my original quote with, "Don't fall in love with your pretty pictures!" Unlike Merlin, we can't become a fish to understand them better, but we can swim in their world.

In this book, I have tried to give you a sense of what it is like to be in the open ocean, both as a permanent inhabitant and as a researcher. Humans are different at sea than they are on land. I already mentioned that we don't think as quickly, maybe because we're working so hard to stand, but there's more to it than that. It's hard for me to put my finger on it, but the best I can say is that people at sea become more of what they are.

Anxious people become more anxious, angry people become angrier, and kind people become kinder, to name just a few traits. I tend to become crasser, more team oriented, and happier at sea.

So, just as we need to understand animals in the context of their habitats, we should ideally try to understand marine biologists in the context of theirs. I've lived in about twenty different houses and apartments in three nations, but when I think of home, I always think of the first ship I sailed on, the RV *Edwin Link*. So, I want to end this book with something I wrote on a ship I loved nearly as much, the RV *Point Sur*. I was chief scientist on the cruise, which left from and returned to Gulfport, Mississippi, in the summer of 2019. The cruise involved trawling, exploring the deep Gulf of Mexico with an ROV, and deploying the Medusa camera system. We got lucky and the Medusa got excellent footage of a giant squid, just 100 miles off the Gulf Coast, so we were in a good mood as we headed for home. The cruise was funded by the National Oceanic and Atmospheric Administration, and it required that we post daily mission logs to its website. As chief scientist, it was my job to get everyone to write one, with me writing the extras. On our journey home, I sat in the corner of the mess, on a bench filled with candy bars, and thought for a while about why it is always so sad to end a cruise. After about twenty minutes, I wrote this. I'm sharing it here because I want to end the book in the voice of who I am when I'm in the world I love:

> Those who go to sea again and again fall in love with the animals, the people, the adventure, and also the ships themselves. Ships are wonderful things, and will tell you much if you listen. Lying in my bunk, the whine of the hydraulic system tells me that the trawl is in the water, the grinding of

the bow thrusters tells me that we're adjusting direction to retrieve our floating camera, and the clatter of dishes tells me that yet another meal is about to be served.

Under it all, though, is the sound of the engine itself, churning day and night to move us, power our lights, and keep us pointed safely into the waves. This fundamental note of our lives for the last two weeks will stop tomorrow morning, and the ship—like Cinderella's carriage—will no longer be a living creature of the sea, but a chunk of metal tied to a commercial dock in Gulfport, Mississippi. I'll be happy to see trees again, to call my wife and daughter, and to walk more than 50 feet in a single direction. But I'll also be sad.

We call it the post-cruise blues, the sadness that lasts for several weeks at the end of any expedition. But why? Yes, the work can be fascinating, but can also be tedious and hard. We also, though well cared for and well fed, lead an offshore life with little sleep, zero privacy, and a ban on certain creature comforts (I would literally kill for a drink). So what are we grieving when we go back to shore, and what brings some of us out here again and again?

It's community. We are 23 out here, in an emptiness that stretches past the horizon in all directions. In these two weeks, our only reminders of the larger world have been two ships and an oil rig at great distance and a thin stream of email. So we only have each other. Many of us were strangers before this cruise, and we come from diverse worlds with diverse beliefs. But we are a community. We have a purpose, and each of us is valued and plays an essential role.

As my friend and shipmate Heather mentioned, we do almost everything as a team. Launching the remotely operated vehicle (ROV), for example, requires ten people scattered throughout the ship. One person's job may be to just hold a line as the ROV is lifted over the side of the deck and into the water. But if this person lets go of this line, the ROV will swing out, and then swing back into the side of the ship with the force of a small truck. Everything we do matters to each other.

As a scientist, I have been trained for decades to use cautious language, to never say that I am certain about anything. But I am certain about this—we want community. We want to be in a group where what we do matters. Many of us spend our lives looking for this, or—if we once had it—finding our way back. And those of us who find it at sea come back again and again. For me, it is an escape from the complex and competitive world on land where it can be hard to find a place.

It's certainly no utopia. We're in a tight space, trying to get a lot done. We're tired, and we often disagree about what should happen next. As chief scientist on this cruise, it is my job to make sure that all the researchers feel that they are being treated fairly, and it is not an easy job (did I mention that drink?). But we care for each other. We catch each other as we stumble on the endlessly shifting deck, we protect each other's heads from hanging hazards, and we nurture each other's dreams (here, wake up, I found the fish that you've been looking for . . .).

Why? Because we have to. We are over a hundred miles from land in a boat that tops out at 11 miles per hour. Imagine driving a sick loved one to an emergency room halfway across your state in a car that can't get out of first gear. If anything happens on a ship, whether an accident or an ugly personal conflict, we have to deal with it. And we do. It turns out that 23 people more or less randomly put on a small boat in the middle of the sea can get along, pull together as a team, and accomplish great things. At any given time we are either exploring the deep with a robot sub controlled from a van bolted to the bow, pulling a large net through the ocean, or setting loose a remote camera through the water column looking for (and finding!) giant squid. We deal with storms, broken equipment, and frustrations of all kinds, but we get it done.

So this is what I will miss when the engine finally stops its churning and we all—after hugs, handshakes, and promises to stay in touch—get into our various vehicles and drive away. I will miss the knowledge that a group of disparate people can come together; treat each other with respect, kindness, and humor; and get the job done. I wish this experience for my university, my town, my adopted country, and for our world, which—like our boat—is just a blip in a sea that goes beyond the horizon in every direction. We are all at sea, and we are all in it together.

# FURTHER READING

Torres, J. J., and T. G. Bailey 2022. *Life in the Open Ocean*. Wiley-Blackwell. (A graduate-level textbook on pelagic biology.)

## Pelagic Animals and Their Ecology

Bigelow, H. B. 1948. *Fishes of the Northwest Atlantic*. 6 vols. New Haven, CT: Sears Foundation for Marine Research.

Bone, Q. 1998. *The Biology of Pelagic Tunicates*. Oxford: Oxford University Press.

David, P. M. 1965. The surface fauna of the ocean. *Endeavour* 24:95–100.

Glasby, C. J., P. A. Hutchings, K. Fauchald, H. Paxton, G. W. Rouse, C. W. Russell, and R. S. Wilson. 2000. Polychaeta. Pp. 1–296 In *Polychaetes and Allies: The Southern Synthesis*, edited by P. L. Beesley, G.J.B. Ross, and Glasby, 1–296. Melbourne: CSIRO.

Haddock, S.H.D., and A. C. Choy. 2024. Life in the midwater: The ecology of deep pelagic animals. *Annual Review of Marine Science* 16:383–416.

Harbison, G. R., L. P. Madin, and N. R. Swanberg. 1978. On the natural history and distribution of oceanic ctenophores. *Deep Sea Research* 25:233–56.

Jereb, P., and C.F.E. Roper, eds. 2016. *Cephalopods of the World. An Annotated and Illustrated Catalogue of Cephalopod Species Known to Date*. 3 vols. Rome: Food and Agriculture Organization of the United Nations.

Kramp, P. L. 1959. *The* Hydromedusae *of the Atlantic Ocean and Adjacent Waters*. Copenhagen: Carlsberg Foundation.

Lalli, C. M., and R. W. Gilmer. 1989. *Pelagic Snails*. Palo Alto, CA: Stanford University Press.

Mayer, A. G. 1910. *Medusae of the World III: The Scyphomedusae*. Washington, DC: Carnegie Institute.

———. 1912. *Ctenophores of the Atlantic Coast of North America*. Washington, DC: Carnegie Institute.

Norman, M., and H. Debelius. 2000. *Cephalopods: A World Guide*. Harxheim, Germany: Conch Books.

Pierrot-Bults, A. C., and K. C. Chidgey. 1988. *Chaetognatha.* Avon, UK: Bath.

Poore, G.C.B. and S. T. Ahyong. 2023. *Decapod Crustaceans: A Guide to Families and Genera of the World.* Melbourne: CSIRO.

Rouse, G. W. 2001. *Polychaetes.* Oxford: Oxford University Press.

Russell, F.R.S. 1953. *The Medusae of the British Isles.* Cambridge: Cambridge University Press.

———. 1970. *The Medusae of the British Isles II: Pelagic Scyphozoa.* Cambridge: Cambridge University Press.

Sutton, T. T., P. A. Hulley, R. Wienerroither, D. Zaera-Perez, and J. R. Paxton. 2020. *Identification Guide to the Mesopelagic Fishes of the Central and Southeast Atlantic Ocean.* Rome: Food and Agriculture Orgainzation of the United Nations.

Totton, A. K. 1965. *A Synopsis of the Siphonophora.* London: British Museum.

Van der Spoel, S. 1976. *Pseudothecosomata, Gymnosomata and Heteropoda.* Utrecht, Netherlands: Bohn, Scheltema and Holkema.

Vinogradov, M. E., A. F. Volkov, and T. N. Semanova. 1996. *Hyperiid Amphipods (Amphipoda, Hyperiidea) of the World Oceans.* Washington, DC: Smithsonian Institution Libraries.

Widder, E. A. 2021. *Below the Edge of Darkness.* New York: Random House.

## Chapter 2: Gravity

Denton, E. J. 1962. Some recently discovered buoyancy mechanisms in animals. *Proceedings of the Royal Society A.* 265:366–70.

Emlet, R. B. 1983. Locomotion, drag, and the rigid skeleton of larval echinoderms. *Biological Bulletin* 164:433–45.

Hardy, A. C. 1956. *The Open Sea, Its Natural History: The World of Plankton.* Boston, MA: Houghton Mifflin.

Mackie, G. O., P. R. Pugh, and J. E. Purcell. 1987. Siphonophore biology. *Advances in Marine Biology* 24:97–262.

Mangum, C. P., and D. W. Wolfe. 1982. The *Nautilus* siphuncle as an ion pump. *Pacific Science* 36:273–82.

Wittenberg, J. B. 1960. The source of carbon monoxide in the float of the Portuguese man-of-war, *Physalia physalis* L. *Journal of Experimental Biology* 37:698–705.

## Chapter 3: Pressure

Kooyman, G. L., M. A. Castellini, and R. W. Davis. 1981. Physiology of diving in marine mammals. *Annual Review of Physiology* 43:343–56.

Pannetoon, W. M. 2013. The mammalian diving response: An enigmatic reflex to preserve life? *Physiology* 28:284–97.

Pelster, B., and P. Sheid. 1992. Countercurrent concentration and gas secretion in the fish swim bladder. *Physiological Zoology* 65:1–16.

Poneganis, P. J., and G. L. Kooyman. 2000. Diving physiology of birds: A history of studies on polar species. *Comparative Biochemistry and Physiology A* 126:143–51.

Schmidt-Nielsen, K. 2008. *Animal Physiology*. 5th ed. Cambridge: Cambridge University Press.

## Chapter 4: Light

Bracken-Grissom, H., D. DeLeo, M. Porter, T. Iwanicki, J. Sickles, and T. M. Frank. 2020. Light organ photosensitivity may suggest a novel role in counterillumination. *Scientific Reports* 10:4485.

Haddock, S.H.D., M. A. Moline, and J. F. Case. 2010. Bioluminescence in the sea. *Annual Review of Mariine Science* 2:443–93.

Herring, P. J., A. K. Campbell, and M. Whitfield, eds. 1990. *Light and Life in the Sea*. Cambridge: Cambridge University Press.

Johnsen, S. 2001. Hidden in plain sight: The ecology and physiology of organismal transparency. *Biological Bulletin* 201:301–38.

Johnsen, S. 2012. *The Optics of Life*. Princeton, NJ: Princeton University Press.

Mobley, C. D. 1994. *Light and Water*. San Diego, CA: Academic Press.

Osborn, K. J., S. H. Haddock, F. Pleijel, L. P. Madin, and G. W. Rouse. 2009. Deep-sea, swimming worms with luminescent "bombs." *Science* 325:964.

Widder, E. A., M. I. Latz, and J. F. Case. 1983. Marine bioluminescence spectra measured with an optical multichannel detection system. *Biological Bulletin* 165:791–810.

## Chapter 5: Motion

Bale, R., M. Hao, A.P.S. Bhalla, and N. Patankar. 2014. Energy efficiency and allometry of movement of swimming and flying animals. *Proceedings of the National Academy of Sciences of the USA* 111:7517–21.

Bone, Q., and E. R. Trueman. 1983. Jet propulsion in salps (Tunicata: Thaliacea). *Journal of Zoology* 201:481–506.

Cohen, J. H., and R. B. Forward Jr. 2009. Zooplankton diel vertical migration: A review of proximate control. *Oceanography and Marine Biology* 47:77–110.

Costello, J. H., S. P. Colin, J. O. Dabiri, B. J. Gemmell, K. N. Lucas, and K. R. Sutherland. 2021. The hydrodynamics of jellyfish swimming. *Annual Review of Marine Science* 13:375–96.

Davenport, J. 1994. How and when do flying fish fly? *Reviews in Fish Biology and Fisheries* 4:184–214.

Frank, T. M., and E. A. Widder. 2002. Effects of a decrease in downwelling irradiance on the daytime vertical distribution patterns of zooplankton and micronekton. *Marine Biology* 140:1181–93.

Gemmel, B. J., J. O. Dabiri, S. P. Colin, J. H. Costello, J. P. Townsend, and K. R. Sutherland. 2021. Cool your jets: Biological jet propulsion in marine invertebrates. *Journal of Experimental Biology* 224. https://doi.org/10.1242/jeb.222083.

Gosline, J. M., and M. E. DeMont. 1985. Jet-propelled swimming in squids. *Scientific American* 252:96–103.

Hays, G. C. 2017. Ocean currents and marine life. *Current Biology* 27:R470–72.

Sutherland, K. R., and L. P. Madin. 2010. Comparative jet wake structure and swimming performance in salps. *Journal of Experimental Biology* 213:2967–75.

Widder, E., and T. M. Frank. 2001. The speed of an isolume: A shrimp's eye view. *Marine Biology* 138:669–77.

## Chapter 6: Food

Choy, C. A., S.H.D. Haddock, and B. H. Robison. 2017. Deep pelagic food web structure as revealed by in situ feeding observations. *Proceedings of the Royal Society B* 284:20172116.

Choy, C. A., E. Portner, M. Iwane, and J. C. Drazen. 2013. The diets of five important mid-trophic mesopelagic fishes of the central North Pacific. *Marine Ecology Progress Series* 492:169–84.

Davis, A. L., K. N. Thomas, F. E. Goetz, B. H. Robison, S. Johnsen, and K. J. Osborn. 2020. Ultra-black camouflage in deep-sea fishes. *Current Biology* 30:3470–76.

Johnsen, S. 2014. Hide and seek in the open sea: Pelagic camouflage and visual countermeasures. *Annual Review of Marine Science* 6:369–92.

Katija, K., C. A. Choy, R. E. Sherlock, A. D. Sherman, and B. H. Robison. 2017. From the surface to the seafloor: How giant larvaceans transport microplastics into the deep sea. *Science Advances* 3:e1700715.

Madin, L. P. 1988. Feeding behavior of tentaculate predators: In situ observations and a conceptual model. *Bulletin of Marine Science* 43:413–29.

Purcell, J. E. 1980. Influence of siphonophore behavior upon their natural diets: Evidence for aggressive mimicry. *Science* 209:1045–47.

Steinberg. D. K., and M. R. Landry. 2017. Zooplankton and the ocean carbon cycle. *Annual Review of Marine Science* 9:413–44.

Steinberg, D. K., B.A.S. van Mooy, K. O. Buesseler, P. W. Boyd, T. Kobari, and D. M. Karl. 2008. Bacterial vs. zooplankton control of sinking particle flux in the ocean's twilight zone. *Limnology and Oceanography* 53:1327–38.

Thompson, T. E., and I. Bennett. 1969. *Physalia* nematocysts: Utilized by mollusks for defense. *Science* 166:1532–33.

Trueblood, L. A., S. Zylinski, B. H. Robison, and B. A. Seibel. 2015. An ethogram of the Humboldt squid *Dosidicus gigas* Orbigny (1835) as observed from remotely operated vehicles. *Behaviour* 152:1911–32.

Tsukamoto, K., and M. J. Miller. 2021. The mysterious feeding ecology of leptocephali: A unique strategy of consuming marine snow materials. *Fisheries Science* 87:11–29.

Wishner, K., E. Durbin, A. Durbin, M. Macaulay, H. Winn, and R. Kenney. 1998. Copepod patches and right whales in the Great South Channel off New England. *Bulletin of Marine Science* 43:825–44.

## Chapter 7: Family

Bracken-Grissom, H., and C. Verela. 2021. A mysterious world revealed: Larval-adult matching of deep-sea shrimps from the Gulf of Mexico. *Diversity* 13:457.

Johnsen, S., and K. J. Lohmann. 2008. Magnetoreception in animals. *Physics Today* 61(3):29–35.

Lohmann, K. J., and C.M.F. Lohmann. 2019. There and back again: Natal homing by magnetic navigation in sea turtles and salmon. *Journal of Experimental Biology* 222 (suppl. 1):jeb184077.

Molacek, J., M. Denny, and J.W.M. Bush. 2012. The fine art of surfacing: Its efficacy in broadcast spawning. *Journal of Theoretical Biology* 294:40–47.

Omori, M. 1974. The biology of pelagic shrimps in the ocean. *Advances in Marine Biology* 12:233–324.

Pineda, J. 2000. Linking larval settlement to larval transport: Assumptions, potentials, and pitfalls. *Oceanography of the Eastern Pacific* 1:84–105.

Robison, B. H., and R. E. Young. 1981. Bioluminescence in pelagic octopods. *Pacific Science* 35:39–44.

## Chapter 8: Community

Hamner, W. M. 1975. Underwater observations of blue-water plankton: Logistics, techniques, and safety procedures for divers at sea. *Limnology and. Oceanography* 20:1045–51.

Hamner, W. M., L. P. Madin, A. L. Alldredge, R. W. Gilmer, and P. P. Hamner. 1975. Underwater observations of gelatinous zooplankton: Sampling problems, feeding biology, and behavior. *Limnology and Oceanography* 20:907–17.

Judkins, H., M. Vecchione, A. Cook, and T. T. Sutton. 2017. Diversity of midwater cephalopods in the northern Gulf of Mexico: Comparison of two collecting methods. *Marine Biodiversity* 47:647–57.

Laval, P. 1980. Hyperiid amphipods and crustacean parasitoids associated with gelatinous zooplankton. *Oceanography and Marine Biology Annual Review* 18:11–56.

Luo, J. Y., B. Grassian, D. Tang, J.-O. Irisson, A. T. Greer, C. M. Guigand, S. McClatchie, and R. K. Cowen. 2014. Environmental drivers of the fine-scale distribution of a gelatinous zooplankton community across a mesoscale front. *Marine Ecology Progress Series* 510:129–49.

McClintock, J. B., and J. Janssen. 1990. Pteropod abduction as a chemical defense in a pelagic Antarctic amphipod. *Nature* 346:462–64.

Pavlov, D. S., and A. O. Kasumyan. 2000. Patterns and mechanisms of schooling behavior in fish: A review. *Journal of Ichthyology* 40 (suppl. 2):S163–231.

# INDEX

A page number in italics refers to a figure.

## A NOTE ON THE TYPE

This book has been composed in Arno, an Old-style serif typeface in the classic Venetian tradition, designed by Robert Slimbach at Adobe.